The Mirage of a Space between Nature and Nurture

The Mirage of a Space between Nature

and Nurture   EVELYN FOX KELLER

DUKE UNIVERSITY PRESS   Durham & London 2010

498112182

© 2010 Duke University Press
All rights reserved

Printed in the United States of America on acid-free paper ∞
Designed by Jennifer Hill
Typeset in C & C Galliard by Keystone Typesetting, Inc.

Library of Congress Cataloging-in-Publication Data appear
on the last printed page of this book.

CONTENTS

ACKNOWLEDGMENTS

I am deeply indebted to the many friends and colleagues who read, thought about, and offered comments on parts or all of this manuscript. Especially, I want to thank Diane Paul, Gregory Radick, Maria Kronfelder, Loup Verlet, Margery Arent, Jehane Kuhn, and Philip Kitcher; their comments were immensely helpful. I also want to thank Emmanuel d'Hombres and Arnaud Pocheville for their invaluable research assistance, and Ned Hall and Rebecca Ann Perry for the figures. It goes without saying, however, that I bear full responsibility for the deficiencies and defects that surely remain.

One of the most striking features of the nature-nurture debate is the frequency with which it leads to two apparently contradictory results: the claim that the debate has finally been resolved (i.e., we now know that the answer is neither nature nor nurture, but both), and the debate's refusal to die. As with the Lernian Hydra, each beheading seems merely to spur the growth of new heads. In the case of the Hydra, Hercules managed to definitively vanquish the beast. I do not pretend to the prowess of a Hercules; my aim is not even to crush the nature-nurture debate. Rather, it is to understand what it is about that debate that causes so much trouble, that so stubbornly resists resolution.

Part of the difficulty comes into view with the first question we must ask: what is the nature-nurture debate about? There is no single answer to this question, for a number of different questions take refuge under its umbrella. Some of the questions express legitimate and meaningful concerns that can in fact be addressed scientifically; others may be legitimate and meaningful, but perhaps not answerable; and still others simply make no sense. I will argue that a major reason we are unable to resolve the nature-nurture debate is that all these different questions are tangled together into an indissoluble knot, making it all but impossible for us to stay clearly focused on a single, well-defined and meaningful question. Furthermore, I will argue that they are so knitted together by chronic ambiguity, uncertainty, and slippage in the very language we use to talk about these issues. And finally, I will suggest that at least some of that ambiguity and uncertainty comes from the language of genetics itself.

For example, we often assume, and indeed often read, that the nature-nurture debate is about sorting out the contributions of nature from those of nurture, and trying to estimate their relative importance. But what exactly is meant by nature and nurture? Sometimes the distinction is between what is inborn and what is acquired after birth; more commonly, it is between genes and environment. Moreover, these terms are themselves ambiguous: what exactly *is* a gene, and what does it do? Even more troublesome is the ambiguity of the term *environment*. Do we mean it to refer to everything other than DNA, to the milieu in which the fertilized ovum develops, or to the factors beyond the organism that affect its development? Finally, there is also the question, contributions *to what*? This, alas, we almost never ask, either as readers or as writers. Yet here we can find what may be the most commonly encountered and the most recalcitrant source of trouble with the entire nature-nurture debate, for what is at issue — the subject of debate — depends critically on our tacit assumptions about how that question is to be answered.

By far the most common assumption — at least in the popular and semipopular literature — is that what is at issue is a comparison of the contributions of nature and nurture to the formation of individual traits. For example, this is the assumption that underlies much of the argument of Matt Ridley's widely read book, *Nature via Nurture* (2003). Ridley's central thesis is that modern genomics has shown us that the nature-nurture debate, as traditionally framed, is premised on a meaningless opposition. He writes:

> The discovery of how genes actually influence human behavior, and how human behavior influences genes, is about to recast the debate entirely. No longer is it nature versus nurture, but nature via nurture. Genes are designed to take their cues from nurture. (2003, 5)

In other words, what matters for development is not so much what genes an organism has, but how and when these genes are expressed — and to be expressed, they need to be activated by environmental stimuli. His take-home message: nature depends upon nurture to be realized.

But in a review of Ridley's book, the evolutionary geneticist H. Allen Orr argues that Ridley misses the main point of the nature-nurture debate. Orr's chief complaint is that Ridley "seems to have the right answer to the wrong question" (2003). What Orr refers to as the "traditional question" of this debate is altogether different from Ridley's concern with how genes respond to experience:

> The first question is statistical. It asks about the percentage of variation in, say, IQ, that arises from inherited differences among individuals (do some parents pass on smart genes to their kids?) versus the percentage that arises from environmental differences (do some parents pass on books to their children?). The second question is mechanistic. It asks about how genes behave within individuals . . . The fact that genes respond to experience is certainly interesting and important . . . But it's the wrong *kind* of fact to settle the nature-nurture debate. (ibid.)

To Orr, the difference between the two questions seems clear, and we might ask (as in fact he does), how so sophisticated a science writer as Matt Ridley could make so elementary a mistake: "why does Ridley reach for the wrong level of analysis, confounding statistics and mechanisms?" Orr suggests that the explanation is as plain as the mistake: Ridley, he writes, "a self-styled champion of 'techno-optimism,' seems to have succumbed to genome hype" (ibid.).

I disagree. What Orr describes as Ridley's confusion between statistics and mechanism is simply too widespread, too difficult for both readers and authors to detect, and too resistant to clarification to be explained by excessive "techno-optimism." The conflation is everywhere, in popular and technical literature alike. It may well be that the distinction seems clear to Orr, but if so, if he himself never slides from one meaning to another, then he is truly an exception.

For another example of the same slippage, and to illustrate its ubiquity, consider the numerous arguments currently being made that invoke epigenetics (rather than genomics) as the crucial new science that resolves the nature-nurture debate. In an article from a recent issue of *Scientific American Mind*, we read:

> Psychologists, psychiatrists and neuroscientists have jousted for years over how much of our behavior is driven by our genes versus the environments in which we grow up and live. Arguments have persisted because there has been little hard evidence to answer basic questions . . . [But today] a field called epigenetics has finally begun to address some of these issues. (Steinberg, 2006)

In a similar vein, an article entitled "Genes or Environment? Epigenetics Sheds Light on Debate," published in a newsletter of the National Institutes of Health, asks:

> Which is more important in shaping who we are and what we will become — our genes or the environment around us? For centuries, people have debated whether nature or nurture decides how we look and act. Now, a field of research called epigenetics is showing that we can't really separate one from the other. (*News in Health*, February 2006)

What is it that makes these claims startling? Not that we have finally found a way to resolve the nature-nurture debate — that claim is about as old as the debate itself. Nor is it that we have finally learned that nature and nurture cannot really be separated. Rather, it is, first, the supposition that that lesson is somehow new, and second, that epigenetics — which I'm all for — is being invoked as the agent of resolution.

Epigenetics is a term that Conrad H. Waddington (1942) coined to refer quite generally to developmental processes (i.e., how we get from genotype to phenotype), and we have known for a long time that such processes involve far more than DNA. In this sense of the term, epigenetics is not a new field. Also not new is the recognition that the various factors involved in development — nucleic acids (DNA and RNA), metabolites, and proteins; nuclear and cytoplasmic factors; genetics and environment — are so deeply intertwined, so profoundly interdependent, as to make any attempt to partition their causal influence simply meaningless. Long before the discovery of DNA, the geneticist Lancelot Hogben was obliged to caution his readers that "genetical science has outgrown the false antithesis between heredity and environment productive of so much futile controversy in the past" (1933, 201).

What is new today involves an altogether different reference of the term *epigenetic*. The "field of research called epigenetics" in the NIH newsletter refers primarily to the discovery that not only are changes in various extra- (or epi-) genetic factors affecting phenotype routinely passed on in cell division, but also such changes can often be transmitted through the generations, despite the fact that they do not involve changes in DNA sequence. Examples of epigenetic inheritance might involve changes in methylation patterns on DNA, or changes in chromatin structure, metabolic requirements, feeding patterns, or even modes of symbolic communication. These alternative "systems of inheritance"[1] are of immense importance to development; they radically change our understanding of inheritance; and they can also have a profound effect on evolution. The important point here, however, is that in themselves, such effects have little if anything to do with the nature-nurture debate as formulated in the examples I cite.

There is surely a muddle here — actually, there are several. There is the muddle about the meaning of *epigenetic* that I have just described; there is another muddle about what kinds of things can be separated, what kinds of questions can be answered; and there is a third muddle about what we take the nature-nurture debate to be about. The casting of the debate as an effort to determine "how much of our behavior is driven by our genes versus the environments in which we grow up and live" poses a question that is not only unanswerable but, as I have already indicated, is actually meaningless. Indeed, we scarcely need the new sciences of genomics and epigenetics to teach us this lesson.

In thinking about development, the causal components belonging to nature are conventionally taken to be the hereditary units that modern biology calls *genes*. The concept of the gene is in considerable disarray these days, but let us for the moment accept the hypothesis that such discrete entities exist as units of inheritance. Just what the components of nurture (environment) might be, I have no idea, but even if we could identify them, we would still have a serious problem. What is the causal role of a gene in the absence of environment? None is clearly the answer. Absent environmental factors, genes have no more power to shape the development of an individual than do environmental factors in the absence of genes. Let us take the simplest possible contemporary definition of a gene on offer: a sequence of DNA nucleotides that codes for a protein.[2] By themselves, such sequences of nucleotides don't do anything: DNA is an inert molecule. What we think of as its causal powers are in fact provided by the cellular complex in which it finds itself. It is this complex that is responsible for both the code that enables a sequence of nucleotides to be translated into a sequence of amino acids, for the replication of DNA, and for the intergenerational fidelity of replication; it is the cellular complex that makes possible all the chemical reactions on which these processes depend. By themselves, the entities we call genes do not act; they do not have agency. Strictly speaking, the very notion of a gene as an autonomous element, as an entity that exists in its own right, is a fiction. In order for a sequence of nucleotides to become what is conventionally called a gene requires that the sequence be embedded in a cellular complex that not only reads, translates, and interprets that sequence, but also defines it, giving it its very meaning.

Not only is it a mistake to think of development in terms of separable causes, but it is also a mistake to think of the development of traits as a product of causal elements interacting with one another. Indeed, the notion of interaction presupposes the existence of entities that are at least ideally separable — i.e., it presupposes an a priori space between component entities — and this is precisely what the character of developmental dynamics precludes. Everything we know about the processes of inheritance and development teaches us that the entanglement of developmental processes is not only immensely intricate, but it is there from the start. From its very beginning, development depends on the complex orchestration of multiple courses of action that involve interactions among many different kinds of elements — including not only preexisting elements (e.g., molecules) but also new elements (e.g., coding sequences) that are formed out of such interactions, temporal sequences of events, dynamical interactions, etc. Compounding the entanglement between genes and environment yet further, as Ridley so clearly explains, biologists now recognize that the development of phenotypic traits is guided not so much by the actual sequence of nucleotides as it is by patterns of gene expression that are themselves products of an immensely complex web of interactions between environmental stimuli (both internal and external to the cell) and the structure, conformation, and nucleotide sequence of the DNA molecule.

Accordingly, as the Swiss primatologist Hans Kummer remarked some years ago — and as Frans de Waal (2002) reminds us — trying to determine how much of a trait is produced by nature and how much by nurture, or how much by genes and how much by environment, is as useless as asking whether the drumming that we hear in the distance is made by the percussionist or his instrument. Richard Lewontin offered another metaphor: "If two men lay bricks to build a wall, we may quite fairly measure their contributions by counting the number laid by each; but if one mixes the mortar and the other lays the bricks, it would be absurd to measure their relative quantitative contributions by measuring the volume of bricks and of mortar" (1974, 401). All of these are intended to make the same point: the effort to separate "under two distinct heads the innumerable elements of which personality [or any other aspect of phenotype] is composed" (as Francis Galton posed the question in 1874) makes no sense; neither personality nor, for that matter, any other feature of an organism is composed of separable elements.

The point is a logical one about which there ought, at least in principle, to be no debate: causes that interact in such ways simply cannot be parsed; it makes no sense to ask how much is due to one and how much to another. As even Steven Pinker (2005) seems at times to recognize, "nature and nurture are not alternatives." This is true whether we read nature as innate and nurture as acquired, or nature as genes and nurture as environment. To drive the point home, my colleague Ned Hall has developed a cartoon version of the philosophical conditions that must be met in order to parse the causes of a phenomenon, a representation he refers to as

1 The Bucket Model.
Adapted from cartoon
by Ned Hall.

Billy

Suzy

Here is a bucket: Billy fills it with 40L of water; then Suzy fills it with 60L
of water. So: 40% of the water in the bucket is due to Billy, 60% to Suzy.

the bucket model. I illustrate with two figures borrowed (and adapted)
from Hall:

In figure 1, innate qualities (or genes) fill the bucket to height x, and
personal (acquired) experiences (or environment) add an amount y. Put
in these terms, it is easy to see that the categories of innate and acquired
cannot be represented in this way, that a kind of logical error has been
committed. We do not have to think of the neonate as a blank slate in
order to appreciate the fact that many, if not all, of the traits we think of as
innate depend upon the acquisition of nourishment, parental care, and
socialization in order to develop. As I will show in a bit, even Locke did
not harbor such a view. The true relation between innate and acquired, or
between genes and environment, is more like that illustrated in figure 2.

Simple, right? But perhaps too simple. If all that was at issue in the
nature-nurture debate was a comparison of the contributions of nature
and nurture to individual development, then both Ridley and the commentators on epigenetics I cited are of course correct: this question is

2 When Causes Interact. Adapted from cartoon by Ned Hall.

But suppose instead that what happened was this: Suzy brought a hose to the bucket; then Billy turned the tap on. Now how much of the water is due to Billy, and how much to Suzy?

Answer: The question no longer makes any sense.

meaningless ("we can't really separate one from the other"), and the debate could indeed be said to be over. But unfortunately, the question of what the nature-nurture debate is about is not so easily settled. As Orr makes abundantly clear, to population geneticists, the debate is not about relative contributions to individual traits, but about contributions to the variation within a population. Still others think of it as being about the relative importance of the contribution of nature and nurture to differences between individuals. Furthermore, not only do different people have different questions in mind, but individual authors (e.g., Ridley) themselves tend unwittingly to vacillate between the various options.

Also, as I have tried briefly to indicate, there is more than a single source of confusion here — in fact, trying to make sense of how arguments about nature and nurture proceed quickly reveals a morass of linguistic and conceptual vegetation grown together in ways that seem to defy untangling. Indeed, it is precisely this morass that is the subject of my book. I want to explore both its conceptual underpinnings and its history. I want to understand how a debate that is so frequently claimed to have been resolved manages, after each such claim, to survive and persist; and how so many readers and authors of manifest intelligence, scientific sophistication, and good faith continue to be enmeshed by the tangle of meanings that make up the nature-nurture debate.

The question that faces us is therefore twofold. First, how did we come to be so entangled, and second, how and why do we persist in making the very same errors, over and over again? The question I am asking is not only how (when, where) a sense of opposition, a versus, came to be inserted between nature (understood either as innate or as genetic) and nurture (understood either as acquired or as environmental), but, more specifically, how did the separation on which opposition is predicated — the *and* on which a *versus* depends — come to be so taken for granted? Whence came the mirage of a space between nature and nurture? How did this illusion become so deeply entrenched in our thinking, and why is it so resistant to dissolution?

The persistence of the nature-nurture debate has been a source of considerable puzzlement to many scholars. Many years ago, the developmental psychologist Daniel S. Lehrman had this to say:

> When opposing groups of intelligent, highly educated, competent scientists continue over many years to disagree, and even to wrangle bitterly, about an issue which they regard as important, it must sooner or later become obvious that the disagreement is not a factual one, and that it cannot be resolved by calling to the attention of the members of one group . . . the existence of new data which will make them see the light . . . If this is, as I believe, the case, we ought to consider the roles played in this disagreement by semantic difficulties arising from concealed differences in the way different people use the same words, or in the way the same people use the same words at different times; [and] by differences in the concepts used by different workers . . . (Lehrman 1970, 18–19)

It is the aim of this book to follow up on Lehrman's suggestion that we look to semantics — to the vicissitudes of the language we use — for an understanding of our impasse. But before proceeding, a word about the peculiar difficulty of my task: in making this analysis, I employ the same language that I critique, and necessarily so. Indeed, I have no other. My description is therefore prone to the very same kinds of slippage I am attempting to diagnose. Even my best efforts do not totally guarantee that the writing of this text has entirely avoided the problem, and for readers who may be less attentive to the problem, the difficulty is that much greater. The net result, I fear, can be somewhat dizzying. Perhaps a schematic outline may help.

Chapter 1 is historical. It focuses on the emergence of a bucket model of the roles of nature and nurture in the formation of traits — i.e., on the assumption already implicit in Francis Galton's catchy phrase, "Nature and Nurture" (1874) that there exist two domains, each separate from the other, waiting to be conjoined. Galton was hardly the first to write about nature and nurture as distinguishable concepts, but he may have been the first to treat them as disjoint. As far as I can tell, such an assumption of mutual exclusivity was not made by earlier writers. For those who used the terms, nurture was rarely, if ever, seen as separable from nature; instead, it was referred to as helping and assisting, or as responding to, nature; nurture was more of a verb than a noun. But those writing after Galton did tend to disjoin the two, increasingly so over time. What is especially noteworthy to me is that the shift in formulation followed directly on the heels of the introduction of a particulate theory of inheritance in the last third of the nineteenth century. Indeed, I argue that this shift was greatly assisted by the arrival of a new way of conceptualizing heredity, and perhaps even dependent upon it.

This remainder of the book brings the story into the modern era. Contemporary biology of development has clearly exposed the assumption in Galton's formulation as meaningless, as not making sense. Indeed, the problem was already evident early in the twentieth century, and in an effort to salvage the questions that had interested Galton, a reformulation of his project was soon provided. The English statistician R. A. Fisher was one of the first who taught us the necessity of making two fundamental distinctions if we wished to address Galton's concerns: we must distinguish first between traits and trait differences, and second between individual and population. Although we may not be able to parse the causal contributions of genetics and environment to individual traits, we *are* able, at least under some circumstances, to statistically parse the causal contributions of differences in genetics and environment to differences in traits averaged over a population.

The following two chapters focus on the nature of these distinctions, the enormous difficulty we seem to have in holding to them, and at least some of the reasons why. In chapter 2, I try to clarify the nature of the distinction between traits and trait differences. I claim that this distinction poses particular difficulty for us when formulated in terms of genetics, and I proceed by locating at least part of this difficulty in the discourse of genetics and its history, briefly indicating at the end of the chapter the ways in which contemporary genetics might offer us a way out of our difficulty. In chapter 3, I focus on the distinction between individual and population that was of such particular importance to Fisher. But here too we seem to have enormous difficulty in keeping the two categories apart, and here too I locate at least part of the problem in language — this time around, in the polysemy of the word *heritability*. In the technical literature of population genetics, heritability was defined as referring to a statistical measure that has meaning only in relation to populations. Unfortunately, however, the word was already in use, but with another, simpler meaning — namely, transmissibility from parents to offspring. The double meaning of heritability has been frequently noted, but, in my view, its role in the continuing confounding of the two meanings, and accordingly of individual and population dynamics — both in the technical and popular literature — has not been adequately pursued. Most scientists tend to underplay the importance of ambiguity (and the equivocation it invites) in scientific argument; they are wont to insist that they know what they mean, and perhaps they do. And certainly, in the immediate context of their experimental practice, where precision is mandatory, scientists generally do know exactly what they mean. But in other contexts (e.g., where words are serving as placeholders for connecting one set of observations with another), there may be considerably less need for, or even less possibility of, precision. In any case, heritability would seem to be a subject about which it is not only easy but virtually inescapable to mean more than one thing at the same time.

My final chapter addresses the question, what is to be done? Is it possible to reformulate the questions at issue in the nature-nurture debate in ways that not only capture what is of primary interest to people, but also, at the same time, lead to meaningful and answerable questions? A central point of this book is to argue that at least part of the explanation for the unreasonable persistence of the debate is to be found in the language of particulate genes, a language that originally developed out of the hope that a particulate theory of inheritance might do for biology what the atomic theory had done for chemistry. For good or bad, biology has turned out to be vastly more complicated, and so has genetics itself. Not surprisingly, however, the language of genetics lags behind, and by doing so, helps to keep alive debates that no longer have meaning — indeed, that may never have had meaning. My hope is that a language that better reflects the contemporary science will be able to help us out of the morass in which we have been bogged down for so long, and help put us on track to providing information relevant to the legitimate concerns that many people have.

ONE Nature and Nurture as Alternatives

Contemporary discussions about nature and nurture cover a broad range of territory, including many concerns that have echoed not only through the ages, but also across cultures. Indeed, we often see the nature-nurture debate traced back to ancient Greek discussions about *physis* (timeless nature, what *is*) and *nomos* (law, custom, culture); about *sumphuton* (innate, internal, or native) and *epikteton* (adventitious, artificial, or acquired); sometimes the debate is even linked to Confucian and neo-Confucian discussions about *hsing* (often translated as nature or innate) and *hsui, yang*, or *chih* (various expressions for nurture, cultivation, or culture). But I am deeply suspicious of such easy assimilations of our own questions with those of people living in different worlds — worlds carved in ways that bear so little resemblance to our own taxonomies — and my misgivings are clearly shared by scholars far more knowledgeable than I about these distant cultures and languages.[1]

In any case, my subject is not nearly so large. Whatever resonance with the concerns of others we might be able to recognize, my subject is limited to the particular form in which we (at least in the English-speaking world) today tend to frame discussions of the roles that nature and nurture play in development. More specifically, my concern is with the tendency to think of nature and nurture as separable and hence as comparable, as forces to which relative strength can be assigned, as suggesting opposability and, accordingly, as inviting debate. Indeed, the very conjunction "nature and nurture" reflects a particular way of conceptualizing the relation between these two concepts: it implies that the two are already disjoint, for only things that are parted can be conjoined. That form of speaking, I claim, is specific to a time and a culture. It came into being at a particular historical moment, and it is sustained by a multitude of cultural, political, economic, and scientific interests characteristic of the particular world in which we live. Yet even so delimited, the subject is still too large. My focus in this book is even more restricted: I am interested in the ways in which that framing is sustained by a particular set of scientific discourses.

In other words, I will not attempt an analysis, or survey or history, of the family of concepts to which we might assimilate our notions of nature and nurture. Nor do I want to disparage the many legitimate and meaningful concerns that such discussions, even when cast as a debate, seem to address. There clearly are issues raised in the debate — even as it is represented in the quotes with which I began — that strike powerful chords for many people; there is something here (perhaps several somethings) that many people clearly do want to know. I will try to address these concerns in my final chapter, but here, in chapter 1, my focus is on the question of how (and when) an understanding of nature as sharply disjoined from our understanding of nurture comes into common parlance. Whence came the mirage of a space between nature and nurture?

Disjoining Nature from Nurture

It used to be claimed that the nature-nurture debate began with Francis Galton, and it certainly was Galton who, with *English Men of Science: Their Nature and Nurture* (1874), put the conjunction into wide circulation. Galton refers to his phrase "nature and nurture" as "a convenient jingle of words" (1874, 12), and indeed it is. But as I've already said, it is also more than that: it is a catchphrase that conjoins two domains on the tacit assumption that they are initially disjoint; it sneaks into our language — as if it had always been there, and as if it were self-evident — the presupposition of disjunction on which conjunction rests. Recently, a fashion has arisen for tracing the phrase "nature and nurture," and the debate with which that phrase is associated, back to Shakespeare, or at least to Prospero in *The Tempest* (1623), who writes off Caliban as uneducable: "a born devil, on whose nature / Nurture will never stick." Some have traced it further back to a monograph on children's education written by an Elizabethan pedagogue, Richard Mulcaster. Mulcaster's words, "Nature makes the boy toward, nurture sees him forward" (1581, 35), are sometimes cited as an early contribution, and perhaps even a beginning, of "the great war." Indeed, if you consult Google on "nature and nurture," you'll find Prospero and Mulcaster repeatedly cited — usually by commentators who in one way or another seek their own fortunes in the supposed war. Steven Pinker, for example, writes: "When Richard Mulcaster referred in 1581 to 'that treasure . . . bestowed on them [children, i.e.] by nature, to be bettered in them by nurture,' he gave the world a euphonious name for an opposition that has been debated ever since" (2004, 5). But in fact, although both Mulcaster and Shakespeare juxtapose the workings of nature and nurture, neither invokes the terms as an explicit conjunction; they do not write of "nature and nurture" as such. Indeed, when we look more closely at what they do write, we can see that their use of the two terms does not invite such a conjunction, for there is no presumption of an a priori disjunction.

Mulcaster's lengthy treatise focuses entirely on the ways in which nurture — meaning here education — can help, assist, and confirm (all his words) that which has been provided by nature, and only occasionally can "procure that which we have not by Nature" (1581, 53). The function of education is to cultivate the course indicated by nature, "to see forward that which nature leans us toward." So, for example, Mulcaster was a strong advocate for women's education, on the ground that nature would never have pointed women toward idleness or smallness of purpose: "young maidens deserve the training because they have that treasure [the ability to learn] . . . bestowed on them by nature, to be bettered in them by nurture" (171). In effect, the very function of nurture is mandated by nature, for it is our nature that makes us nurturable, or trainable. Thus, not only does nurture not stand here in opposition to nature, but the two cannot even be separated. For Mulcaster and Shakespeare, the aim of education is to "nurture nature";[2] a nurture that is set in opposition to nature—directed toward ends that are too alien to a child's nature—is useless, even meaningless. On a creature that is born too ill disposed, like Caliban, "Nurture can never stick" (*The Tempest*, act 4, scene 1).

John Locke did not explicitly juxtapose the terms nature and nurture, but he did write extensively about the relation between education and innate dispositions. And since he has so often been represented as a champion of sensationalism, an enemy of the doctrine of innate ideas, and the bête noire of the "blank slate," it is worth noting, first, that innate ideas are not the same as innate dispositions and, second, how very dialectical his views on the latter actually were. Note, for example, his repeated advice to parents and educators that they

> well study their [children's] natures and aptitudes, and see by often trials what turn they easily take, and what becomes them; observe what their native stock is, how it may be improv'd, and what it is fit for . . . For in many cases, all that we can do, or should aim at, is, to make the best of what nature has given, to prevent the vices and faults to which such a constitution is most inclin'd, and give it all the advantages it is capable of. Every one's natural genius should be carry'd as far as it could; but to attempt the putting another upon him, will be but labour in vain; and what is so plaister'd on, will at best sit but untowardly, and have always hanging to it the ungracefulness of constraint and affectation. (1692, part I, section 66)

And later in the same text, he urges them to "consider which way the natural make of his mind inclines him. Some men by the unalterable frame of their constitutions, are stout, others timorous, some confident, others modest, tractable, or obstinate, curious or careless, quick or slow. There are not more differences in men's faces, and the outward lineaments of their bodies, than there are in the makes and tempers of their minds" (ibid., part I, section 101). And still later, "observe your son's temper . . . See what are his predominate passions and prevailing inclinations" (ibid., part I, section 102).[3]

I'm not certain just when a sense of separability or oppositionality between nature and nurture appeared in the English-speaking world, but it seems clear that there is no such sense either in Mulcaster's or Shakespeare's use of these terms, or in Locke's arguments for the importance of education. Similarly, it seems not to have been present in the eighteenth century. Of the relationship between the natural and the artificial in that century, Lorraine Daston writes that the categories "could overlap, and the boundary between them was often blurred" (1992, 223). And when she adds that "the relationship between the natural and the artificial in general, and between nature and education (or upbringing) in particular, was never so starkly complementary as that between nature and nurture" (ibid.), it is clear that she is referring primarily to the concepts of nature and nurture that became predominant in the last third of the nineteenth century. Although the view of nurture as "powerless to change nature" (224) may not have been entirely alien to earlier writers, as Waller (2001) argues, I submit that it was only after Darwin and Galton that such a view became commonplace. Similarly, I suggest it is mainly late-nineteenth-century arguments that Daston has in mind when she writes: "attempts to subvert the dictates of nineteenth-century nature were deemed futile; attempts to subvert those of eighteenth-century nature, perverse" (1992, 224).

I am not even sure that I can find a sense of separability or oppositionality in the writings of John Stuart Mill to which Charles Darwin takes such exception. Let us look at Darwin's complaint:

> Mr. J. S. Mill speaks, in his celebrated work, 'Utilitarianism,' (1864, pp. 45, 46), of the social feelings as a "powerful natural sentiment," and as "the natural basis of sentiment for utilitarian morality." Again he says, "Like the other acquired capacities above referred to, the moral faculty, if not a part of our nature, is a natural out-growth from it; capable, like

them, in a certain small degree of springing up spontaneously." But in opposition to all this, he also remarks, "if, as in my own belief, the moral feelings are not innate, but acquired, they are not for that reason less natural." It is with hesitation that I venture to differ at all from so profound a thinker, but it can hardly be disputed that the social feelings are instinctive or innate in the lower animals; and why should they not be so in man? (*Descent* [2nd ed., 1879], part 4, note 240)

What exactly is Darwin's concern here? Although Mill would appear to be posing an opposition between innate and acquired, he immediately follows by undermining that disjunction: "they are not for that reason less natural." The moral faculty, though acquired, may not be part of our nature, but it is "a natural out-growth from it." Indeed, although not included in Darwin's excerpt, Mill's text continues with an even more explicit expression of his denial of an opposition between the two: "It is natural to man to speak, to reason, to build cities, to cultivate the ground, though these are acquired facilities" (*Utilitarianism*, 1861, 230). But Darwin ignores this sequitur, focusing instead on the disjunction that Mill rhetorically denies while at the same time expressing a view that would substantially affirm it. Social feeling would fall in the same category as man's ability to speak: natural even though requiring experience and education for its development.[4] Perhaps a hint of the source of Darwin's concern might be found in his particular understanding of "transmitted qualities." In his concluding sentence, he writes: "The ignoring of all transmitted mental qualities will, as it seems to me, be hereafter judged as a most serious blemish in the works of Mr. Mill" (*Descent* [2nd ed., 1879], part 4, note 240).[5]

The Turning Point

The first explicit use of the terms nature and nurture as an unambiguous disjunction — as clear alternatives, even if not necessarily in opposition — that I have been able to find comes from the very same year, and it appears not in Darwin's writings, but in the text (to which I've already referred) written by his cousin, Francis Galton.[6] But before turning to the writings of Galton, I want to suggest that there is already in Darwin's dissent from Mill a clear hint of the turn that Galton makes explicit. This turn, I claim, is rooted in changing conceptions of heredity, and in accord with these changes, with the new alignment between innate and hereditary then taking place. I am not persuaded that there is anything in Mill's writings to indicate such an equation between innate and hereditary, still less to support an equation between nature and heredity. In fact, in many of Mill's remarks, *hereditary* refers primarily to the inheritance of property or title; as for most writers of his time, the noun *heredity* was not yet part of his usual vocabulary.[7]

But in Darwin's thinking,[8] such an alignment is clearly already in view; furthermore, he himself contributed greatly to bringing this realignment into common parlance. Indeed, translating the recently coined French word *hérédité*, Spencer and Darwin were probably the first to introduce *heredity* into the English language, and they did so in the early 1860s. Carlos López Beltrán has argued that the term brought with it a new ontological commitment to the material concreteness of whatever it was that lay behind hereditary processes, and I think he is right.[9] The crucial point is that, for Darwin, heredity is both nominalized and interiorized. Not the law, not civil or church code, not custom, and not theological prescription, but the body is now the vehicle of inheritance. Here is where the concept of heredity begins to take on its modern meaning, referring not to external inheritance, but to the transmission of something biological, of some substance that resided inside the body. Accordingly, when Darwin writes of the "transmission" of qualities, physical or mental, he is already referring to the passing on of an internal substance. Indeed, I suggest that it is precisely the simultaneous internalization and substantiation of heredity that makes the alignment of the notion of inborn or innate with that of heredity seem, as it were, so natural.

During the remaining part of the century, with the considerable help of Francis Galton, and later of biometricians like Karl Pearson, the translation of innate to hereditary appears ever more explicitly and ever more frequently; by the end of the century, at least within the biological community, the equation has been consolidated, and the terms of the nature-nurture debate decisively transformed. Innate has given way to heredity, and acquired to environment.[10] What this shift does is to displace Mill's differentiation between innate and acquired — understood as a distinction marked by an event in time (birth) — by a distinction that cuts along a different axis, a division between internal and external, and concomitantly between the kinds of substances belonging to these different spaces, rather than between influences acting over different periods of time. Furthermore, this change to a new set of axes corresponds to much more than a change of variable: it introduces a distinction of a different kind. Where before we had an effectively nominal distinction, one readily compatible with the construal of influences after birth "nurturing" those that came before, now we have a concrete, material division, a division between different kinds of things found in different spaces.

But there is also another point that I need to make. The new conception of heredity, especially as formulated by Galton, had one additional component, a feature that I believe was especially crucial to casting the respective influences of nature and nurture — now understood in terms of internal and external factors, heredity and environment — into a relation of competition and opposition. For Darwin writing in the 1860s, and certainly by 1868,[11] what was internal to the body and transmitted from generation to generation was cast as a particulate something, a substance composed of submicroscopic elements (Darwin called them "gemmules," or "invisible characters")[12] that "represented" or were "responsible for" the various aspects of the organism.[13] As I will show, Darwin's gemmules take on an even more particulate (or atomic) character for Galton than they did for Darwin, becoming not only more independent, but also invariant. Where Darwin held that these particles might be shaped by the experience of an organism, beginning with Galton, they were reconceived as fixed entities that were passed on from generation to generation without change.[14] The relevance of this difference to the causal separation of nature and nurture is that although the effects of malleable (or soft) hereditary particles might be regarded as separable from the effects of "nurture" within a single generation, over the course of many generations, their influence would become hopelessly entangled with the influence of experience (or "nurture").

In fact, Galton's own writings are our best source here. In the text already cited, he offers as a reason "the phrase 'nature and nurture' is a convenient jingle of words" the fact that "it separates under two distinct heads the innumerable elements of which personality is composed" (1874, 12). Here certainly is a hard disjunction between the causal effects of nature and nurture, and it is explicit. What exactly is the disjunction between? On the one hand, it is between "nature," initially defined by Galton according to the convention of his time as "all that a man brings with himself into the world," and "nurture," defined, equally conventionally, as "every influence that affects him after his birth" — in other words, between pre- and postnatal influences (i.e., as a temporal divide). But on the other hand, and in the same breath, the difference between nature and nurture is characterized as a substantive disjunction — i.e., as a divide between the two different kinds of "elements of which personality is composed." One set of elements belongs to "nature," and the other, to "nurture."

To be sure, this reformulation does not give us the means to make such a causal partition in practice; it only claims that partition's feasibility. Furthermore, Galton goes on to disavow any implication that either of the terms "implies any theory; natural gifts may or may not be hereditary; nurture does not especially consist of food, clothing, education or tradition, but it includes all these and similar influences whether known or unknown" (1874, 12). He even acknowledges that neither is self-sufficient. Nevertheless, he also claims that "the distinction is clear: the one produces the infant such as it actually is, including its latent faculties of growth of body and mind; the other affords the environment amid which the growth takes place," and he insists that "when nature and nurture compete for supremacy . . . , the former proves the stronger." Moreover, his professed agnosticism about the character of the components of nature is scarcely evident in his actual practice; in virtually all his investigations, the elements belonging to nature — what he refers to in his preface as "pre-efficients," a term that he defines as "all that has gone into the making of eminent men" (vi) — are assumed to be elements of heredity.

In Galton's mind, such a distinction[15] was not only clear; it was also necessary to make explicit. He was well aware of the criticism mounted against his emphasis on the importance of heredity, especially in the development of genius, but he believed that separating what belongs to nature from what belongs to nurture would make a conclusive argument possible; indeed, for him, the argument *depended* on "drawing the necessary distinction" (1876a, 391). The difficulty was how to do so in practice. Seeking a "method by which it would be possible to weigh in just scales the respective effects of nature and nurture," he seized on the idea that his program could be implemented through the study of twins. The history of twins, he writes, "affords the means of distinguishing between the effects of tendencies received at birth, and of those that were imposed by the circumstances of their after lives; in other words, between the effects of nature and nurture."

Particulate Genetics

The ability to make such a separation would certainly have been useful to Galton's argument (whether or not it was necessary), but the question remains, why would he have supposed it to be possible? What was it that made the formulation of nature and nurture as separable causal domains — the formulation of a disjunction that, to my knowledge, had not before (either in the scientific or the philosophical literature) been so sharply posed, and was certainly not posed by Mulcaster, Locke, or Mill — not only possible, but suddenly so clear to Galton? The answer that I propose lies precisely in the dual transformation of the terms *nature* and *nurture* — first from their conventional association with pre- and postnatal influences (as indicated by Galton's own definition of the terms) to their conceptualization as two distinguishable substances, and second to their configuring as different ensembles of distinct and separable causal elements, each making its own contributions to the formation (or composition) of human personality.

Galton (1871) was greatly influenced by Darwin's "provisional theory of Pangenesis" (Darwin [1868]). Although he ultimately disputed, and effectively refuted, the part of Darwin's theory that suggested the susceptibility of gemmules to environmental modification, Galton retained his cousin's commitment to the existence of hereditary particles. In fact, in Galton's hands, hereditary particles became noticeably more particulate — i.e., more discrete and independent — than they were for Darwin. For example, they were not (at least, not for the most part) subject to any actual blending (1872). Indeed, as Galton later argued, even the "blending in stature is due to its being the aggregate of the quasi-independent inheritance of many separate parts" (1889, 139).[16]

What he took from Darwin without qualification was the hypothesis that "the body consists of a multitude of 'organic units,' each of which possesses its own proper attributes, and is to a certain extent independent of all the others" (Galton 1876b, 329, quoting the 1868 edition of Darwin). Furthermore, Galton insisted that the implications of this hypothesis "must lie at the foundation of the science of heredity." What are those implications? As Galton saw them, they are:

1　Each organic unit "has a separate origin, or germ."
2　The "stirp" — defined as "the sum-total of the germs, gemmules, or whatever they may be called, which are to be found, according to every theory of organic units, in the newly fertilized ovum" (1876b, 330) — contains a host of germs, much greater in number and variety than the organic units of the bodily structure that is about to be derived from them; so that comparatively few individuals out of the host of germs, achieve development.
3　The undeveloped germs remain vital though "latent," and are passed on to the next generation.
4　The organization of organic units depends "on the mutual affinities and repulsions of the separate germs."[17]

That germs must be separate from one another — like the separateness of atoms and molecules[18] — is indeed a recurrent theme in Galton's writings. In his view, the only alternative hypothesis was that of "a general plastic force"—a hypothesis he likened to "other mystic conceptions current in the early stages of many branches of physical science, all of which yielded to molecular views, as knowledge increased" (1876b, 332).[19]

In other words, Galton presumed a theory of development (a theory he first elaborated in 1872) in which separate, particulate (and invariant[20]) germs develop into organic units, which subsequently combine and arrange themselves to produce body parts, more or less independently of variations in the environment. At least for body parts, "nature"—here understood as an ensemble of distinct elements, each of which has its own causative powers—seemed sufficient to account for the specificity of these parts. But to defend such a view of the development of personality traits was clearly more problematic, and Galton was from the start explicit about his desire to demonstrate this fact "more pointedly" than had been done before (1865, 157). His conviction that mental qualities are as much under the influence of internal hereditary factors as are physical qualities preceded even Darwin's arguments to this effect. Darwin's influence on Galton is unmistakable, but on this subject the influence was mutual. Darwin's growing commitment to the importance of "transmitted mental qualities" seems to have been greatly strengthened by his reading of Galton, and the affinity between the notion of "transmitted mental qualities" and Galton's claims that personality (like body parts) is "composed of innumerable and separable elements" (1874, 12) is unmistakable. Yet, as Galton was also obliged to acknowledge, "mental habits in mature life are the creatures of social discipline, as well as of inborn aptitudes" (1865, 320). The challenge, as he saw it, was to sort the one from the other, and especially "to ascertain what is due to the latter alone."

But can that be done? If, as I claim, such a separation cannot be effected unless a bucketlike model of genes and environment applies, with the different causal factors assumed to both act and be transmitted independently, then, according to our understanding of development, we would have to say that Galton's challenge makes no sense. But perhaps a bucket model might be made to fit with his model of development. If so, we might be better able to understand the strength of his commitment to particulate inheritance (as opposed to the hypothesis of "general plastic forces"). In any case, once the notion of "elements" was extended to environmental forces as well—a move Galton makes without comment in the original formulation of his aim in *English Men of Science* (1874)—his radically particulate view of development and inheritance does clearly support the idea that the various elements can be sorted into two distinct camps. In other words, "nature" and "nurture" have become alternatives. And once sorted, the temptation to think of them as if in competition may have been hard to resist.

The Question of Eugenics

So far, I have said nothing about Galton's advocacy of eugenics, and readers will surely be wondering what relation, if any, Galton's insistence on the particulate nature of heredity—and relatedly, on the separability of nature and nurture—had to his social agenda. I therefore conclude this chapter with a few brief remarks on that relation. We know that Galton developed his position on all three of these issues (particulate inheritance, heredity of mental traits, and eugenics) concurrently, and also that in his mind, as well as in the minds of many of his readers, each issue and each claim was just a piece of a larger whole. Part of the point (if not the main point) of a theory of heredity, to Galton, was to facilitate a program of what he originally called "Hereditary Improvement" (the title of his first popular call for the establishment of a society aimed at the improvement of the human stock, published in *Fraser's Magazine* in 1873) and what, ten years later, he dubbed "eugenics" (1883). His aim in "Hereditary Improvement" was to demonstrate "that it is feasible to improve the race of man by a system which shall be perfectly in accordance with the moral sense of the present time" (1873, 116). Moreover, he was also persuaded, and sought to persuade his readers, that the need for conscious race betterment (taking for granted that a reference to race was automatically a reference to nature) would become evident to anyone who learned the truths of heredity.

Galton may have coined the term *eugenics*, but, as a number of historians have clearly shown,[21] he was hardly the first to put forth hereditarian arguments for a program of race improvement. Yet his commitment to such a program seems to have been particularly strong. In his case, anxiety about the decline of his race (or the decline of his nation), and a felt need to take positive steps to reverse this decline, went hand in hand with his interest in heredity as a science, and more specifically, with his vision of a particulate and quantitative genetics. Similarly, by the time of his 1873 paper, published two years before his theory of heredity, he had already come to take the disjunction of nature and nurture for granted. Indeed, his argument in this paper proceeds on the basis of that assumption:

> There is nothing in what I am about to say that shall underrate the sterling value of nurture, including all kinds of sanitary improvements; . . . nevertheless, I look upon race as far more important than nurture. Race has a double effect, it creates better and more intelligent individuals, and these become more competent than their predecessors to make laws and customs, whose effects shall favourably react on their own health and on the nurture of their children. Constitutional stamina, strength, intelligence, and moral qualities cling to a breed, say of dogs, notwithstanding many generations of careless nurture, while careful nurture, unaided by selection, can do little more to an inferior breed than eradicate disease and make it good of its kind. (1873, 116–17)

As Pearson, in his biography of Galton, later paraphrased the argument, "the best environment will not free mankind from weaklings, they can only be 'bred out'" (1914, 118). What worries Galton is whether his own race (usually referring to his fellow Englishmen) will be able to withstand the challenges he anticipates in the future, and he is especially preoccupied with what he sees as an insufficiency in the supply of talent and genius for the years to come — hence his focus on "Hereditary Talent and Character" (1865), "Hereditary Genius" (1869), and *English Men of Science* (1874).[22] But his 1873 article focuses on the practices that threaten to "spoil the race":

> The first is, the free power of bequeathing wealth, which interferes with the salutary action of natural selection, by preserving the wealthy, and

by encouraging marriage on grounds quite independent of personal qualities; and the second is the centralising tendency of our civilisation, which attracts the abler men to towns, where the discouragement to marry is great, and where marriage is comparatively unproductive of descendants who reach adult life. (1873, 117).

Finally, he concludes:

It is no absurdity to expect . . . that while helpfulness to the weak, and sympathy with the suffering, is the natural form of outpouring of a merciful and kindly heart, yet that the highest action of all is to provide a vigorous, national life, and that one practical and effective way in which individuals of feeble constitution can show mercy to their kind is by celibacy, lest they should bring beings into existence whose race is predoomed to destruction by the laws of Nature. It may come to be avowed as a paramount duty, . . . by endeavouring to breed out feeble constitutions, and petty and ignoble instincts, and to breed in those which are vigorous and noble and social. (ibid., 118–20)

I cite these passages at some length, partly to show both the strength of Galton's views, and partly to show their consonance with his more technical arguments about heredity. At the same time, it seems evident to me that, although they may have come together in both Galton's mind and his work, there is no logical mandate linking the separability of nature from nurture with the aims of eugenics. Not only were his notions about particulate inheritance not shared by earlier advocates of race improvement, they were also not necessarily shared by those who came later. Similarly, there is no necessary eugenics mandate in the doctrine of particulate inheritance. Galton's own commitment to particularity was undoubtedly inspired by Darwin, but the ways in which he put this commitment to work in relation to his concerns about the future of his race — his efforts to show ("more pointedly" than had been done before) the importance of biological heredity, especially for mental traits — were of his own creation. Many of his eugenics sympathizers (most notably, Karl Pearson) embraced his entire program, but, I repeat, it was not necessary to do so, and not all of them did. Jane Hume Clapperton, for example, a prominent socialist, feminist, and eugenicist of her day, apparently felt no need to assume a disjunctive relation between nature and nurture. To her, the importance of eugenics followed easily from a more traditional view, one in which even though nature and nurture are not in conflict, one can nevertheless readily see that "the power of nurture is limited. It can direct the forces of nature, but it can not alter the intrinsic quality of the raw material which nature provides" (Clapperton 1885, 365–66).

Historically, Pearson's influence was of course infinitely greater than Clapperton's, and Pearson did embrace Galton's social and scientific agenda as a package. In particular, the need for "race improvement" was for Pearson, just as it was for Galton, an important backdrop to his own efforts to sort the effects of nature versus nurture — or, as he later came to put it, the effects of heredity versus environment.[23] I suggest, however, that — for the particular purpose of understanding the origin and persistence of the belief in the separability of nature and nurture (my original question) — the issue of eugenics might logically, even if not historically, be put aside.

In any case, it is to the persistence of this illusion — and the place of Galton's particular vision of the relation between nature and nurture in the modern imagination — that I now need to turn. Galton may not have been in a position to perceive the defects of his formulation, but the modern reader clearly is. Today, it is widely accepted by contemporary biologists and lay readers alike — as Daniel Dennett puts it, "everyone knows"[24] — that genes and environment must interact to produce any biological trait, that nature (understood as heredity) and nurture (understood as environment) are not alternatives. And yet. And yet, the image of separable ingredients continues to exert a surprisingly strong hold on our imagination, even long after we have learned better. Although "everyone knows" it not to be true, many authors continue to argue as if nature and nurture were (or, at the very least, could be regarded as) separable and clearly distinguishable causes of development.

TWO

Changing the Question to One That Does
Make Sense — From Trait to Trait Difference

G alton's question (how much of our personality is due to nature
and how much to nurture?) may have been badly posed, but is
there not something in it of interest to many people that *could* — perhaps in
some other way — be legitimately asked? Galton worked hard to formulate
meaningful measures of the relative importance of heredity, and his efforts
helped launch the science of biometrics. Karl Pearson continued Galton's
efforts, and, building on his work, founded the discipline of mathematical
statistics. But it was probably the English mathematician R. A. Fisher
(1890–1962) who contributed most to reformulating Galton's question.
Fisher devoted his life to trying to distinguish genetic from environmental
influences, and he was clearly aware of the difficulties involved. Like Pearson, he shared many of Galton's concerns; perhaps especially, he too felt
that a science of eugenics was much needed, both socially and scientifically. In 1911, while still a student, he helped form the Cambridge
University Eugenics Society (together with John Maynard Keynes, R. C.
Punnett, and Charles Darwin's son Horace). And a few years later, in an
effort to save what was meaningful in Galton's quest, Fisher (1918) published a reformulation of that quest in a paper that was to be enormously
important in shaping the future of population genetics. Here he clearly
recognized — indeed, he was one of the first to so clearly recognize — the
point that if it were to be realizable, Galton's hope of sorting genetic from

environmental influences would need to be recast in two important ways. First, it was necessary to reformulate the question of causation in terms of trait differences rather than in terms of traits per se, and second, it was necessary to turn from the analysis of heredity in individual lineages to the analysis of heredity in populations. Only if we ask a statistical question about the relative contributions of variations in genetics and in environment to our differences from each other — rather than their relative contributions to the processes that make us what we are — would we have a question that makes sense, and furthermore, one that we might be able to answer.

First Questions First

Asking about contributions to phenotypic differences rather than to the making of a particular phenotype still leaves a potential ambiguity, and thus we need to ask, what exactly do we mean by difference? In one meaning, difference refers to pairs of particular individuals. Returning to the cartoon characters of my colleague, Ned Hall, we might want to know what makes Suzy taller than Billy.

Historically, however, the interest in human differences that has repeatedly manifested itself is of another, far more general, sort. We want to know not simply what makes Suzy different from Billy, but also what makes people like us different from people like them. That is, we have a tendency to sort people into demarcated groups, with the express intention of comparing them to one another. For example, we like to ask what makes women different from men, whites different from blacks, Europeans different from Asians, achievers different from nonachievers, etc. And as I've already mentioned, in order to address questions of this sort, a second move is required. The interest in what makes classes of people different from one another requires us to shift our focus not simply from trait to trait difference, but also from individuals to populations. Of course, in doing so, we must also decide what the population of interest is. For example, are people like us Caucasians, females, septagenarians, Americans,

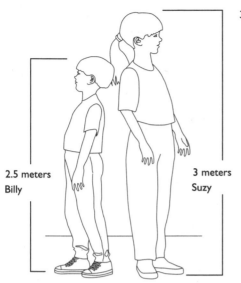

3 Comparing Traits. Adapted from cartoon by Ned Hall.

2.5 meters
Billy

3 meters
Suzy

academics, hikers, or troublemakers? Needless to say, claims that address differences between whatever groups we happen to focus on are those that arouse the most intense and bitterest controversies of all.

In 2005, Lawrence Summers, then president of Harvard University, set off a firestorm with his speculation that innate differences between the sexes might account for women's underrepresentation in math and science. His remarks were ill advised for all sorts of reasons, scientific as well as political, and Summers was to live to regret them. But there are two reasons why I am not here going to address — at least, not directly — the question of why those claims (or the many other similar claims with which we are daily confronted in the mass media) are problematic. My first reason is that the most immediately obvious problem with such claims — namely, the conspicuous failure of the cases investigated to meet the basic requirement of noninteraction between nature and nurture that is required for any parsing of causal contributions to trait differences — has already been eloquently addressed by many others (for example, Lewontin 1974; Block 1996; Bateson 2004). The central issue in these discussions is whether or not the interdependence of genes and environment in development necessarily undermines any measurement involved in the effort to partition the statistical variance.[1]

Perhaps more important, however, is my second reason for not focusing on these controversial claims: the issues I do want to examine here are, I claim, anterior. These are the underlying muddles we get into whenever these topics arise. They concern the difficulties we have, first, in making the distinctions that are required for turning the nature-nurture debate into potentially meaningful questions, and second, once those distinctions are made, the extraordinary difficulties that virtually all of us — scientists and lay readers alike — have in keeping to them.

Both historically and scientifically, the two shifts that Fisher's reformulation requires — from trait to trait difference, and from individual to population — have tended to occur together. But for the sake of trying to illuminate the conceptual aspects of the difficulties we have with them, I want to tease the two apart, and to examine each one separately. I will do so in order. In this chapter, I focus on the distinction between trait and trait difference, formulated not as a difference between populations but as a difference between individuals, deferring my discussion of the distinction between individual and population to chapter 3. In both chapters, I argue that a good part of the difficulty we have in maintaining the necessary shifts is to be found in the slippage of meaning, or polysemy, of our basic vocabulary. When the same term can be used to refer to two different concepts, it should be obvious that any attempt to distinguish one from the other becomes hopelessly confounded. Yet this, I claim, is precisely the situation for both of the distinctions with which we are here dealing.

Parsing the Causes of Trait Differences

What makes Suzy taller than Billy, Amy shyer than Mary, Richard more agile than Peter? There is nothing wrong with such questions: they are perfectly legitimate and, indeed, seem natural to ask. Sometimes it might even be possible to answer them. It turns out that, for quite straightforward technical reasons, questions like this about human behavioral traits can almost never be answered, but my point here is a different one. It is to call attention to the fact that such questions are of a different kind from those we might ask about what makes Suzy grow to almost six feet in height, or about what is responsible for Amy's stuttering, or for Richard's ability to leap five-foot hurdles.

In fact, this is precisely the point of Hans Kummer's analogy to drums and drumming. It is useless to ask whether the drumming that we hear in the distance is made by the percussionist or his instrument because each of the two variables on which the sound — the percussionist's performance (x) and the resonance of the instrument (y) — is influenced by the other in ways that simply do not permit separation. Yet if we hear two different sounds of drumming in the distance, we can ask and perhaps determine whether the difference between the two sounds is caused by a difference in drummers or by a difference in drums, or even how much of the net difference in sound is caused by the former and how much by the latter — provided, of course, the drummer's performance does not depend on the drum being used.

The questions differ not only in what is being asked, the circumstances in which they make sense, and the methods by which they might be answered, but also in the interests that give rise to them. Certainly, the new question — about the causes of difference — is one in which many people have considerable interest. We want to know what accounts for the difference in sound between different musical performances; even more, we want to know what makes people different. In the first case, the reason for our interest seems obvious: we want to find ways of making music sound better. In the second case, the link to practical intervention — to finding ways of making better (taller, smarter, more sociable) people — seems more problematic, both technically and politically. Although it may have seemed straightforward to Galton, the acknowledgment of such a link, especially in the case of the effect of genetic differences on behavioral traits, has in more recent times become so fraught with unacceptable political implications as to be generally avoided.[2]

In any case, it turns out that there is a rather stringent condition that must obtain for questions of this sort to be answerable (or indeed, for them to be well posed), and, using the terms of our example, it is this: that there be no interaction between the quality of the drummer's performance and the quality of the drum. But should this condition apply — i.e., should the performance of the drummer be precisely the same on any drum — it would be logically possible to parse the causal contributions of a change in drummers and a change in drums to the net difference in sound. Equally, if there were no interaction between changes in genetic and environmental (nongenetic) factors, it would be logically possible to parse the causal contributions of alterations in each of these factors to a change in phenotype. Were such conditions to be satisfied, we would be able to say, for instance, that 40 percent of the difference between the two sounds we heard was due to the change in drums, and 60 percent was due to the change in drummers. Or that 45 percent of the difference in height between Suzy and Billy was due to the difference between their genotypes, and 55 percent was due to nongenetic differences (e.g., nutrition) between the two. This can be important information, yet it still does not allow us to say anything about the proportion of genetic to nongenetic influence in the formation of the trait.

The Elusiveness of Clarity

It may seem unduly cumbersome to fully specify the ways in which the question has been altered in each and every iteration, but failing to do so just makes it that much easier to lose sight of the difference. Even Wikipedia, although clearly registering the fact that the question has been changed, takes a shortcut: it casts the nature-nurture debate in terms that explicitly mark the shift of focus to trait differences, but not the shift from the respective roles of nature and nurture to the respective roles of *changes* in nature and nurture in causing trait differences. Thus it reports:

> *Nature vs Nurture* is a shorthand expression for debates about the *relative importance* of an individual's innate qualities ("nature") versus per-

sonal experiences ("nurture") in determining or causing individual differences in physical and behavioral traits.[3]

Furthermore, even for apparently meticulous formulations, a good deal of care is warranted on the part of us as readers. Kummer may be clear about the difference, but the rest of us are not. In fact, it has proven extraordinarily difficult for all of us, both as readers and as writers, to keep hold of the distinctions that have been made here. Indeed, even some of our most eminent philosophers, at times, write as if they too have lost sight of the difference. For example, Elliott Sober has given some of the clearest expositions of the confusions I have been discussing, yet even his lucid prose sometimes invites slippage. For example, consider his response to the claim that the entanglement of genes and environment makes it impossible to parse the causes of development. He writes: "Does this mean it makes no sense to ask whether genes or environment are more important causes of the resulting phenotype? No. It just means that we must pose the question in a different way" (Sober 2000, 357). My point is that we are not simply reposing a question, but substantively changing it.

On first reading, I assumed Sober's reference to "the question" reflected no more than a momentary lapse in his normally meticulous prose, but a subsequent paper made me begin to suspect that there may be more at issue here. "Separating Nature and Nurture" refers to the "attempt to disentangle the contributions of nature and nurture"—e.g., when breeders "think about what makes corn grow tall" (Sober 2001, 47). Sober's main point in this paper is that "the concepts of phenotypic, environmental, and genetic variance apply to human beings just as much as they do to cows and corn" (48), despite the fact that, for humans, significant technical difficulties impede our ability to sort out these different variances from one another. In other words, the difficulties we face are not just political, but also scientific. I agree. What troubles me about Sober's account is the failure to specify exactly what contributions are being disentangled, and to what they are contributing. Sober does not intend his question to collapse into Galton's—his disentanglement is of the contributions of nature and nurture (more precisely, of differences in nature and nurture) to phenotypic differences, not to traits themselves. But for the most part, this distinction goes unmarked, leaving the reader free to interpret as he or she wishes. Indeed, the same ambiguity also haunts Sober's effort to explicate the limitations of the associated statistical analyses. For example, he writes: "The analysis of variance permits one to infer *how much* a cause contributes to an outcome without understanding *how* the cause manages to have its effects" (italics in original). But the problem is not with the distinction between how and how much, but rather with the distinction between two different kinds of outcome. One (the question of how) refers to an individual trait, and the other (the question of how much), to a difference between traits.

A similar slighting of the difference between the two questions can be seen in C. Kenneth Waters's otherwise illuminating analysis of the relation between actual and potential causes (2007). Indeed, following James Woodward (2003),[4] Waters interprets the relation "*x* causes *y*" as meaning "a difference in *x* causes a difference in *y*." This interpretation enables him to slide easily from statements about genes' "causing" traits to statements about gene differences' causing trait differences, and vice versa.

The demands of experimental practice amply justify Woodward's (and Waters's) focus on manipulability, and I totally sympathize. Furthermore, where the outcome depends on only one variable, the relation between the two kinds of statements is relatively easy to see. Put in terms of elementary calculus, if we represent the dependent variable y as a function of x (the independent variable), we would say that y can be obtained from df/dx by integration — provided one knows the derivative df/dx for all values of x. But where more than one variable is at issue (as, e.g., in the case of the developmental processes that give rise to a trait such as height, where a vast number of factors participate), and we wish to assess the relative importance of each such factor (or variable), a huge problem arises. First, we must identify all the variables involved (itself a far from trivial task), and second, we must assess their independence, for it is only among independent variables that we can begin to partition the different causal influences, assigning independent weight to each. For the development of height, the list of variables surely includes both genetic and environmental factors, but we can rarely, if ever, be confident that we have identified all the factors that might be involved. Furthermore, few if any of these causal factors operate independently; indeed, their very demarcation is, like the demarcation of traits themselves, often quite arbitrary.

Even situations where all the variables have been clearly demarcated and enumerated — even when, say, there are only two variables at stake — there is still a serious problem. While it is certainly legitimate to infer that a and b causally participate in the production of y from observations showing that both changes in a and changes in b cause change in y, unless a and b are the only causal factors, and unless they act independently, such observations do not permit us to answer questions like, how much of y is due to a and how much to b? They might enable one to ask (and answer), how much of the change in y is due to the change in a, and how much to the change in b? But the problem with which I am concerned arises when we attempt to infer from the answers to that question (*how much*) an answer to the earlier question of *how*, because in general, this is not possible. With Waters's interpretation of causal statements of the form "x causes y" as meaning "a difference in x causes a difference in y," it becomes all but impossible to mark, or even to recognize, the distinction between the two questions, and correspondingly easy to think that by answering the second, one has also answered the first.

The reader might wonder why I am belaboring the issue. How much does the difference between the two questions really matter? After all, knowing how changes in each variable alter the outcome must surely tell us something about the role each variable plays in producing the outcome — if it did not, why would scientists spend so much effort studying the effect of specific changes? Indeed, such measurements seem to lie at the very heart of the experimental method, and, as I have already acknowledged, a finite effect of a change in variable does identify that variable as a causal factor. The two questions are clearly related, but in systems as complex as the developing organism, that relation can be extremely opaque.

This is a lesson classical geneticists knew well. At least when pressed, they readily acknowledged that the study of phenotypic differences was of limited use in trying to tease out the influence of genotype on phenotype.

But as I discuss below, phenotypic differences were all they had access to, and the habit of conflating the two questions—of attempting to infer a causal relation between genes and traits from analyses of trait differences[5]—may well have begun with the limits of their methodology. But this habit has obvious dangers: just consider, e.g., the folly of attempting to understand the causal dynamics of vision by studying all the ways in which blindness (an extreme example of phenotypic difference) can be induced.

Molecular geneticists are in many ways far more fortunate than their classical precursors. They have direct access to the entity that we generally accept as being of primary importance to the development of traits— namely, the DNA. They can, for example, directly induce specific changes in the DNA and then study the effect such changes have on phenotype. But despite such direct access to at least part of the causal substrate of development, anyone expecting to be able to infer the causal dynamics of DNA from studying the effect of differences in that molecule on phenotype is subject to a few surprises. For example, the effect of changing a variable that is itself known to be causally important to the production of the phenotypic end product may be (in fact, often is) reduced or erased by a system of buffering that is built into the dynamical networks mediating between genotype and phenotype. Indeed, when such effects (or rather, the absence of effects) were first observed, geneticists greeted these results with surprise and consternation. More recently, however, such insensitivity to changes in contributing variables has come to be recognized as the hallmark of systems designed to be robust in the face of common fluctuations. In such systems, the difference effected by a change in variable is no guide at all to the causal importance of that variable.

Of course, genetics is not the only context in which questions about the causal dynamics of a process are conflated with questions about the effect of perturbations on that process, yet I strongly suspect that the tendency is particularly acute here, and for reasons that lie deep in the language of that science. Indeed, when talking about traits, the conclusion that it is possible to decompose a difference in phenotype, even when one can say nothing about the composition of the phenotype, itself seems to contradict basic sense. Imagine a situation in which we were able to conclude that drinking four extra glasses of milk a day throughout her childhood was responsible for 70 percent of the difference in height between Suzy and her non-milk-drinking sister Mary, and still not be able to say anything about the relative contributions of genetic and nongenetic influences to Suzy's (or Mary's) height. That seems altogether counterintuitive, the result of some sleight of hand. But I suggest that at least part of the reason it seems so counterintuitive is that an actual sleight of hand — slipping from talk about differential effects to talk of underlying causes — has become so routine in discussions of genes and environment as to go unnoticed, even to the point of having become part of much conventional reasoning about genetics.

For an example of this sleight of hand, I turn to yet another philosopher of science, Neven Sesardic. Sesardic is known for his extensive critique of the arguments that Lewontin, Jencks, Block, and others have made, and that I have here taken as standard. Much of Sesardic's argumentation depends on close reading of the works that Lewontin et al. take to task, but Sesardic's own formulations unfortunately escape comparable scrutiny. Although his arguments center on the concept of heritability that I discuss only in chapter 3, I include them here because of their illustrative conflations between traits and trait differences on the one hand and between the causal impact of genes and gene differences on the other. For example, he writes: "The idea that heritability reflects the causal strength *of genetic influences on phenotypic differences* has been consistently opposed by a number of authors. It has been said, e.g., . . . that it is dubious whether a clear meaning can be given to *'genetic determination of traits'* (Burian); that inferences about genetic determination of traits should be 'disavowed once and for all' (Kitcher); . . . and so forth" (1993, 399, italics added). Notice that the difference between "the causal strength of genetic influences on phenotypic differences" and the "genetic determination of traits" is completely elided here. In a similar vein, Sesardic elsewhere asks: "Does the fact that a given phenotypic trait is heritable entail that it is genetic (i.e., that the differences in that trait are due to genetic differences)?" (2003, 1002). This question clearly implies that we can equate the claim that a phenotypic trait is genetic with the claim that differences in that trait are due to genetic differences. Sesardic's recent book (2005) on the subject of genetic causation is permeated by the same slippage. Throughout his writing, Sesardic sets himself apart from virtually all other philosophers of science in insisting that measures of the causal factors giving rise to trait difference can and do inform our understanding of the causal dynamics of individual trait development. Unfortunately, however, his argument rests on a routine confounding of the two questions. Thus, in response to David Lykken's claim that "it is meaningless to ask whether Isaac Newton's genius was more due to his genes or his environment, as meaningless as asking whether the area of a rectangle is more due to its length or its width" (Lykken 1998, quoted in Sesardic 2005, 55), Sesardic writes: "Contrary to what [Lykken] says, however, it makes perfect sense to inquire whether Newton's extraordinary contributions were more due to his above-average intellect or to an above-average stimulating intellectual environment." And in response to what he refers to the "mistaken conclusion" by the Nuffield Council on Bioethics — namely, that "it is vital to understand that neither [the broad or narrow] concept of heritability allows us to conclude anything about the role of heredity in the development of a characteristic in an individual" — he writes: "On the contrary, if the broad heritability of a trait is high, this does tell us that any individual's phenotypic divergence from the mean is probably more caused by a nonstandard genetic influence than by a non-typical environment."[6]

What I find curious in all these discussions is that, although the conflation is not restricted to genetics, it does seem both less common and less automatic when talking, say, about drums and drummers. In the latter context, the conclusion that it is possible to decompose a phenotypic difference even when one can say nothing about the relative strength of the causal inputs to the trait itself seems considerably less counterintuitive; indeed, I doubt that anyone would be tempted to infer anything at all about the causal contribution of a drum to the sound that was produced from my hypothetical conclusion that 40 percent of the difference between sounds was due to a difference between drums. So what is it that especially confounds our thinking when we turn our attention to the subject of genetic influences?

Genes as Difference Makers

My claim that an important source of confusion lies deep in the language of genetics requires us to look at the evolution of that science. What precisely is genetics, as a branch of biology, about? How we answer this question of course depends on when, and where, we look. But for classical genetics, especially the paradigmatic school of T. H. Morgan, genetics was about tracking the transmission patterns of units called genes. What was a gene? No one knew, but notwithstanding this ignorance, a gene was assumed to be a unit that could be identified by the appearance of mutants in wild-type populations. That is, a phenotypic difference in some trait (a mutant) was taken to reflect a difference (a mutation) in some underlying gene associated with that trait. But to argue for such an identification between phenotypic difference and underlying gene requires a two-step move. First, change (a mutation) in some underlying entity (the hypothetical gene) is inferred from the appearance of differences in particular phenotypic traits (e.g., blue eye, bent wing, narrow leaf), and second, the existence and identity of the gene itself is inferred from the inference of a mutation. In other words, the classical gene was on the one hand identified *by* the appearance of phenotypic differences (mutants), and on the other hand simultaneously identified *with* the changes (mutations) that were assumed to be responsible for the mutants. Thus, the first map of the mutations thought to be responsible for the observed phenotypic differences in *Drosophila* was called not a map of mutations, but a genetic map, a map of genes.

This is the sense in which the classical gene is often said to be a "difference maker" (see, e.g., Sterelny and Griffiths 1999; Moss 2003). But a gene was taken not only to be a difference maker; it was also assumed to be a trait maker. It was both the entity responsible for the difference observed, and (at least implicitly) the entity responsible for the trait which has undergone a change — i.e., the trait in which a difference has been observed.[7]

We might say, then, that a certain confounding of traits and trait differences was built into the science of genetics from the very beginning;

moreover, we might argue that such a confounding inhered in the logic of their method. The occurrence (and frequency) of trait differences was what geneticists had observational access to: by examining phenotypes, they could detect phenotypic differences — which, in turn, were taken as indicative of changes in some underlying, internal entity. Through breeding, the locus of such changes could be mapped. As Horace Freeland Judson has observed: "In 1913, Alfred Sturtevant, a member of Thomas Hunt Morgan's fly group at Columbia University, drew the first genetic map — 'The linear arrangement of six sex-linked factors in *Drosophila*, as shown by their mode of association.' Ever since, the map of the genes has been, in fact, the map of gene defects" (2001, 769). Similarly, much of the same slippage can be identified in the use of the term "gene" in what Lenny Moss (now along with numerous others) calls "Gene-P," or "phenotypic gene," where the unit of interest is clearly a difference maker. The actual referent of Gene-P is thus not a gene but a genetic mutation.[8] Wilhelm Johannsen, the man to whom we owe the word *gene*, was himself clearly worried about this problem when he asked: "Is the whole of Mendelism perhaps nothing but an establishment of very many chromosomal irregularities, disturbances or diseases of enormously practical and theoretical importance but without deeper value for an understanding of the 'normal' constitution of natural biotypes?" (1923, quoted in Moss 2003, 62).

It is hard to imagine that the early slippage was entirely accidental. To think of genes simply as difference makers would have been to detract from the power of the gene concept. Mapping difference makers and tracking their assortment through reproduction may have been all that the techniques of classical genetics could do, but the aims of these scientists were larger. What made genes interesting in the first place was their presumed power to mold and form — in a word, their presumed power to act. *Gene action* was the term used for the process by which genes exerted their power in the development of characters or traits. But to illuminate the nature of this process (the developmental process), studies of trait differences alone would not suffice. In fact, neither mapping of the locus of the factors (difference makers) presumed to be responsible for such differences nor analysis of their emergence and intergenerational patterns of transmission taught us anything about the causal dynamics of the developmental process by which the traits themselves came to be. As John Dupré puts it, "classical genetics was about invisible features that could trigger different developmental outcomes, but not about the causal explanation of these outcomes" (2006, 118). Furthermore, classical geneticists were for the most part well aware of this distinction. Nevertheless, the easy slide between genes as difference makers and genes as trait makers perpetuated the illusion (as widespread among geneticists as it was among their readers) that an increased understanding of the effects of gene differences would enhance our understanding of what it is that the entities called genes actually do.

What Is a Disease?

A similar confounding of the etiology of traits with that of trait differences pervades virtually all of the current literature of medical genetics. Indeed, the very notion of a disease as an individual trait — in the sense that brown eyes, say, is a trait — already incorporates this confusion. We may commonly speak of an individual as "having a disease," much as he or she might have brown eyes, but as many writers have long understood, disease is a state only in relation to another state that has already been established as normal.[9] In his inquiry into the scientific rules for distinguishing the normal from the pathological, written more than a hundred years ago, Émile Durkheim stressed that "a trait can only be characterized as pathological in relation to a given species" ([1894], 4; my translation) — in other words, in relation to a standard of normality or state of health which is, itself, inextricably confounded with the norm of a species. He continued: "One cannot even conceive, without contradiction, a species that could, by itself and in virtue of its fundamental constitution, be irremediably sick. [The species] is the norm par excellence and, accordingly, can harbor within itself nothing of the abnormal."[10]

It is true that the French philosopher of science Georges Canguilhem, following Kurt Goldstein, made valiant efforts to internalize the diagnostic criteria of pathology, locating them within the individual,[11] but for all his efforts, by far the most common understanding of disease has continued to rely on comparison (or contrast) with a preestablished conception of the normal. English-language dictionaries routinely define disease as a relational state: it is a dis-ease, "an abnormality of the body or mind" (Wikipedia); "a departure from the state of health" (*Oxford English Dictionary*); "a deviation from or interruption of the normal structure or function of a part, organ, or system of the body" (*Dorland's Illustrated Medical Dictionary*); "an interruption, cessation, or disorder of a body, system, or organ structure or function" (MediLexicon). And indeed, French dictionaries do the same. In virtually every dictionary I have consulted, a *maladie* is an *altération de l'état de santé*. In French as in English, an animal can be said to have brown eyes whether or not a comparison with other animals is at hand, but it cannot, without such a comparison, be said to have a disease.

Like medicine, genetics too might be said to be a comparative science. Comparing organisms with differing phenotypes, along with attempting to correlate these phenotypic differences with corresponding genetic differences (mutations), has been the bread and butter of geneticists from the earliest days of that science. But genetics aims beyond comparative judgments, seeking an understanding of the developmental dynamics. And, as I have tried to show, its language invites us to lose sight of the complex moves — first, attributing the cause of a phenotypic difference to a genetic mutation; second, assuming that the presence of a mutation automatically signals the presence of a gene; and third, attributing responsibility for the trait in question to the gene in which the mutation is assumed to have occurred — that are routinely made in effecting this shift from comparative to individual. It seems no accident, therefore, that the adoption of a lexicon of illness referring to disease as an individual attribute comes with the emergence of a medical science grounded in genetics.[12] Medical genetics is a contemporary science aimed at identifying the causes of "inborn errors of metabolism" (i.e., of disease), and it is a direct heir to the more general efforts of classical genetics to associate mutations with particular phenotypic differences. And the main difference between the two kinds of efforts is that, where classical geneticists sought the causes of phenotypic difference in mutations that were for a long time largely hypothetical, the genetic defects that medical geneticists identify as the causes of a disease state can be clearly specified. The main similarity, on the other hand, is that they are subject to exactly the same sorts of conflation. A disease, we need to remember, is not a trait but a trait difference, and the causal factors involved in producing the former are not the same as those involved in producing the latter.

Like the classical geneticists before them, today's medical geneticists are wont to slide from the identification of a genetic difference (a mutation, a departure from a presumed normal genome) associated with a disease state to assertions of having identified a gene responsible for the disease in question. Where classical geneticists relied on "gene maps" to identify the gene presumed to be involved, today medical geneticists are more likely to rely on analyses of nucleotide sequences. They may continue to talk about *genes*, but it is the DNA itself that has become the focus of investigation. The direct object of much actual research is the identification of any change (or defect) in the nucleotide sequence that is correlated with the expression of the disease. Despite widespread talk of "disease genes," or "disease-causing genes," the information these analyses provide is about the causal implications of mutations, and not about the causal role of genes. Indeed, the notion of a gene "causing" a disease (or even of a particular sequence "causing" a disease) has exactly the same status as the notion of a gene "causing" a phenotypic difference.

Furthermore, and perhaps more importantly, the identification of one or more anomalies in the DNA sequence may or may not point to a defect in a particular gene (however that term is defined). Indeed, for diagnostic purposes (at least for diagnoses based on genetic tests), the attempt to correlate a disease state with an underlying gene may in many if not most cases be largely irrelevant. Contemporary medical genetic diagnoses rely on the identification of aberrant or anomalous sequences, and not on the causal pathways such anomalies may disrupt. Also, the anomalies may be anywhere in the genome — in fact, only rarely are they found in protein-coding sequences (that is, in the segments of DNA usually associated with genes).

For most of us, the crucial questions are, can the identification of such sequences be useful in the treatment or prevention of disease, and if so, how? Most immediately (and perhaps most obviously, even if generally left unstated), such information can be used to promote selective abortion. But if we are interested in therapeutic medicine, we need more than simple correlation between aberrant sequence and aberrant phenotype. It is true that the early days of the Human Genome Project brought the promise that in time we would be able simply to replace defective sequences with normal ones (gene therapy). However, that hope has so far failed to materialize, and at least one of the reasons is that the relation between DNA sequence and phenotype turned out to be far more complicated than originally expected. As to the possibility of other kinds of treatment or prevention in a particular individual carrying the aberrant sequence, this depends on understanding something about the biological function that has been disrupted by the identified change in sequence. Such a quest takes us beyond the analysis of phenotypic differences induced by mutant forms. Indeed, it requires an altogether different kind of analysis, almost always one of a far more difficult nature.

There are, however, some relatively simple exceptions. Phenylketonuria (PKU) is one, and it is probably the most celebrated case of therapeutic intervention in the history of medical genetics. It is everyone's canonical example, mine as well. PKU is a disorder (now recognized as genetic) associated with a range of disabling symptoms, including mental retardation, and it is caused by the inability of the body to properly metabolize the essential amino acid phenylalanine. A major breakthrough in the treatment of this disease came with the recognition that its symptoms can be significantly alleviated if the affected individual adheres to a carefully monitored low-phenylalanine diet for his or her entire life.[13] However, the development of a strategy to treat PKU had nothing to do with either the identification or the mapping of the gene(s) or genetic sequence(s) involved. Today we know that this disorder is caused by one or more mutations in the gene encoding the enzyme that breaks down phenylalanine (to date, as many as 400 such mutations have been identified). In point of historical fact, however, neither our understanding of the (at least proximal) cause of this disease (Jarvis 1937) nor the development of a therapeutic intervention (Bickel et al. 1954) depended on any sort of genetic analysis. It is now believed that the phenotypic expression of PKU (the disabling symptoms) are the direct consequence of the accumulation both of high levels of phenylalanine and of toxic intermediates resulting from its faulty metabolism, but we learned this from direct biochemical analysis; ordinary medical observation showed how dreadful the symptoms can be. Furthermore, we have now acquired the ability to precisely characterize many of the mutations responsible for the absence of the necessary enzyme, and doing so has certainly been instructive. The bottom line, however, is that, thus far, that ability has not significantly added to the possibilities of therapeutic intervention.

Of course, it need not have happened that way. Identification, location, and characterization of the guilty mutation(s) could have provided the starting point for a research program that, over time, led to an understanding of the disease and possibilities for treatment. But whatever the sequence of events, there is no way in which the genetics of difference alone could have produced that understanding. What was required for a causal analysis of the disease — and, hence, for the possibility of therapeutic intervention — was a biochemical analysis of the metabolic pathway that had gone astray. That analysis *might* have begun with the identification of a particular gene, but if it had, what would have been required is an understanding of the gene's downstream effects, of the particular role actually played in development by the gene in which the mutation had occurred.

What Genes Do

The challenge of attributing causal function to genes (rather than to gene differences) — of understanding what genes do — has plagued genetics from the onset, and it was not until the advent of molecular biology that it seemed possible to meet this challenge. What does a gene do? Defined as a discrete stretch of nucleotides, a gene was said to "make," or code for, a protein. But even with that insight, it was one thing to be able to track the causal effect of a genetic difference (mutation, or change in nucleotide sequence) on a particular trait, and quite another to track the causal influence of the gene itself (or indeed of the protein it was said to "make") on the development of that trait.

To be sure, enormous progress has been made since the early days of molecular biology, and we now know a great deal more about the ways in which the process of development makes use of an organism's DNA. But what we have learned has not so much answered earlier questions as it has transformed them. We have learned, for instance, that the causal interactions between DNA, proteins, and trait development are so entangled, so dynamic, and so dependent on context that the very question of what genes *do* no longer makes much sense. Indeed, biologists are no longer confident that it is possible to provide an unambiguous answer to the question of what a gene *is*. The particulate gene is a concept that has become increasingly ambiguous and unstable, and some scientists have begun to argue that the concept has outlived its productive prime.

DNA, by contrast, is a concrete molecule — an entity that can be isolated and analyzed, identified by its distinctive physical and chemical properties, and shown to consist of particular sequences of nucleotides. We not only know what DNA is, but with every passing day, we learn more about the exceedingly complex and the multifaceted role it plays in the cellular economy. It is true that many authors continue to refer to genes, but I suspect that this is largely due to the lack of a better terminology. In any case, continuing reference to "genes" does not obscure the fact that the early notion of clearly identifiable, particulate units of inheritance — which not only can be associated with particular traits, but also serve as agents whose actions produce those traits — has become hopelessly confounded by what we have learned about the intricacies of genetic processes. Furthermore, recent experimental focus has shifted away from the structural composition of DNA to the variety of sequences on DNA that can be made available for (or blocked from) transcription — in other words, the focus is now on gene expression. Finally, and relatedly, it has become evident that nucleotide sequences are used not only to provide transcripts for protein synthesis, but also for multilevel systems of regulation at the level of transcription, translation, and posttranslational dynamics. None of this need impede our ability to correlate differences in sequence with phenotypic differences, but it does give us a picture of such an immensely complex causal dynamic between DNA, RNA, and protein molecules as to definitely put to rest all hopes of a simple parsing of causal factors. Because of this, today's biologists are far less likely than their predecessors were to attribute causal agency either to genes or to DNA itself—recognizing that, however crucial the role of DNA in development and evolution, by itself, DNA doesn't *do* anything. It does not *make* a trait; it does not even encode a program for development. Rather, it is more accurate to think of DNA as a standing resource on which a cell can draw for survival and reproduction, a resource it can deploy in many different ways, a resource so rich as to enable the cell to respond to its changing environment with immense subtlety and variety. As a resource, DNA is indispensable; it can even be said to be a primary resource. But a cell's DNA is always and necessarily embedded in an immensely complex and entangled system of interacting resources that are, collectively, what give rise to the development of traits. Not surprisingly, the causal dynamics of the process by which development unfolds are also complex and entangled, involving causal influences that extend upward, downward, and sideways.

THREE

From Individuals to Populations

The principal point of the preceding chapter was to highlight the distinction between asking about the causal contributions to a trait difference between two individuals and asking about the causal contributions to the formation of that trait, and to try to understand why it is so difficult for us to keep this distinction in view. As scientists have clearly recognized for the better part of a century, analysis of the relative importance of different causal factors is possible only in relation to the first question, and only under conditions in which the causal factors are not confounded by one another. In genetics, these conditions are hard to achieve. To drive this last point home, I want to return to my canonical example of PKU and consider two individuals with a measurable difference in IQ, a clear genetic difference (one has a mutation in the relevant sequence, and the other does not), and a clear difference in environment (one is raised on a normal diet, the other on a controlled, low-phenylalanine diet). Now let us ask, how much of the observed difference in IQ is due to the genetic difference, and how much to the difference in diet? This question is well posed, but alas, it cannot be answered. The reason is straightforward: individuals with the mutation react to the difference in diet very differently from those who do not. We can answer questions concerning other variables — variables that act independently of one another — but not about variables that are so entangled. For example, we should be able to answer the question of how much of the difference in observed IQ between two individuals — one with and the other without the mutation, one with and the other without a daily regimen of exercise — is due to genetic and how much to environmental differences, because here the causal factors presumably do act independently of one another.

Yet even where a difference between two individuals can, at least in principle, be clearly analyzed, in practice, it is far from easy to do so. Few studies in human biology can be so carefully controlled (ensuring no other genetic differences and no other environmental differences obtain), and those that can (be they on humans or on other organisms) are almost certain to be studies not of individuals but of populations. Indeed, it is population differences that seem to be of the greatest interest to many people, including sponsors of research, researchers, and the general public. For instance, Galton was interested in the inheritance of intellectual giftedness in the general population; farmers want to know how to boost average crop yields; educators want to know about the effect, on average, that different educational strategies have (again, on average) on students' test scores; behavioral geneticists want to know about statistical variations among groups identified in relation to particular traits (e.g., schizophrenia); and the general public seems to be keenly interested in tracking the causes of observable differences of any kind between men and women or blacks and whites, or of differences in achievement between two groups of children.

Questions about differences between groups require a different kind of analysis than do questions about differences between individuals. A successful analysis of the difference in height between Suzi and Billy still does not tell us about what makes one group of people (say, men), on average taller than another group (say, women) — no more than the analysis of the difference in two distinct drumming sounds tells us about variation in the sounds of drumming in general. For group differences, the question we need to ask is, how much of the variation we hear in the sound of drums is due to variation in drummers, and how much is due to variation in drums? And to answer this question, we must turn to the statistical analysis of populations. Which is precisely how Fisher reformulated Galton's question, and he was clearly aware of the importance of such a shift. Introducing his very first paper on the subject, Fisher warned that while "it is desirable on the one hand that the elementary ideas . . . should be clearly understood; and easily expressed in ordinary language," nonetheless "loose phrases about the 'percentage of causation' which obscure the essential distinction between the individual and the population, should be carefully avoided" (1918, 399–400). Perhaps he even had Galton in mind.

More recent authors have not fared much better. The difficulty in maintaining the essential distinction between individual and population persists, and it may even have been compounded. When authors write about sorting genetic from environmental contributions to the development of traits, it is not only the distinction between trait and trait difference that has been tacitly erased, but also the distinction between individual and population. In this chapter, I will argue that, like the first erasure, the second also is rampant — among behavioral geneticists, evolutionary psychologists, and journalists alike, in popular and semipopular literature as well as in technical journals. Its persistence long after Fisher's warning is puzzling, and my aim is to suggest that once again, at least part of the reason for that persistence lies in our language. That is, our difficulty in maintaining this conceptual distinction is sustained, if not caused, by the words we use.

Words, Words, Words

The culprit here is not the word *gene*, but the closely related words *heritable* and *heritability*. The latter is a term that has become widely used in behavioral genetics in the effort to get at questions about the relative importance of nature and nurture, and it is used in this literature with a very specific technical meaning that was first introduced in 1936. But unfortunately, as has already been noted numerous times, the word itself was not new. Even then, it was in common use, but with a quite ordinary meaning: it referred simply to the quality of being passed on from parent to offspring — i.e., the quality of being inheritable, or just heritable. Although a number of others have noted the double meaning of the term "heritability,"[1] in my view, the implications of that ambiguity have not been adequately pursued. I will argue, first, that slippage between the two meanings of the term is chronic in both the technical and the popular literature, and second, that this slippage is a primary source of our continuing difficulty in keeping claims about populations apart from claims about inheritance per se.

I begin my argument by calling upon technical and ordinary dictionaries to define these terms; I then try to clarify what technical measures of heritability do and do not imply and compare these implications with the conclusions that are so frequently, albeit mistakenly, drawn, arguing that the logic of the mistaken inference is grounded in a slippage between ordinary and technical meanings of the term.

I start by asking, what does it means to say that a trait is heritable? This word would seem to be at least as straightforward as *genetic*. That is, to be heritable is to be capable of being passed on through the generations, of being inherited. And indeed, that is what ordinary dictionaries say. Here is a sample of definitions of *heritable* from familiar, commonly used, dictionaries:

"Hereditary." (*Random House Webster's* 2001; *Webster's International* 2002)

"Capable of being inherited, inheritable [civil law sense]." (*Oxford English Dictionary* 1989; *Shorter Oxford* 1993; *Random House Webster's* 2001; *Webster's* 2002; *Oxford American Dictionary* 2005)

"Transmissible from parents to offspring, capable of inheriting." (*Concise Oxford* 1984)

"Naturally transmissible or transmitted from parent to offspring, hereditary." (*Oxford English Dictionary* 1989; *American Heritage Dictionary* 1991; *Oxford American Dictionary* 2005)

Technical dictionaries, by contrast, seem not to include the word, suggesting that it is not regarded as having, or requiring, a special technical definition: its meaning is self-evident.

But look at what happens when we turn to the cognate noun, heritability. In ordinary dictionaries, the transition to cognate noun (at least since 1832) appears to be straightforward; the noun's definition remains just as it was before the introduction of a technical meaning. In ordinary diction-aries, *heritability* is still defined as:

"The quality of being heritable, or capable of being inherited." (*Oxford English Dictionary* 1989)

"The quality or state of being heritable." (*Webster's* 2002)

But not so in technical dictionaries, and indeed, not so in much of the biological literature more generally. Here, *heritability* has another, techni-cal, meaning that refers not to individual lineages but to populations. The term is defined by Wikipedia and by technical dictionaries as:

"The proportion of phenotypic variation in a population that is attribu-table to genetic variation among individuals. Variation among indi-viduals may be due to genetic and/or environmental factors. Herita-bility analyses estimate the relative contributions of differences in genetic and non-genetic factors to the total phenotypic variance[2] in a population." (*Wikipedia* [accessed March 15, 2007])

"The ratio of genetically caused variability to total variability of a charac-ter in a population. A measure of correlation between genotype and phenotype." (*Glossary of Genetics*)

"The proportion of phenotypic variation attributable to genetic differ-ences between individuals." (*Dictionary of Biology*)

Measuring Heritability

By most accounts, the technical meaning of *heritability* was introduced in 1936 by Jay Laurence Lush, an animal breeder. Lush used the word to refer not to the quality of being inherited from parent to offspring, but to a statistical quantity associated with the ratio of genetic variation to phenotypic variation within a specified population of organisms. Inter-estingly, however, Lush felt no need to give an explicit definition, and as A. Earl Bell pointed out years ago (1977, 297), he showed no awareness that he was coining a new term. Apparently, he took it for granted that existing definitions would cover his new usage. Indeed, Lush's response to a query about this matter sent him by Bell in the 1970s is well worth reading in its entirety (ibid.); here, however, I simply note his acknowledgment that he had considered the term to be roughly synonymous with what Darwin referred to as "the strength of the hereditary tendency" in the fifth edition of *Origin of the Species* (1869, 62).

Lush's technical measure itself had at least two variants, and about this he was explicit; he was also careful to distinguish between the two forms of the term. He called the first measure *narrow heritability* and the second *broad heritability*, and here too his terminology has prevailed. Narrow heritability — typically designated as h — is the measure most commonly used in agriculture: it is the proportion of total phenotypic variation that is due to the additive variation in genes (i.e., what is left when one leaves out any variation due to genetic interactions, either between genes or between alleles), and it is a good indicator of the responsiveness of the population to selection; furthermore, it is a quantity readily obtained from the correlation between parent and offspring phenotype. Broad heritability — typically designated as H — is the proportion of total phenotypic variation that is due to the total genetic variation, including that coming from interactions, and this is the measure more commonly used in behavioral genetics. This quantity may be more intuitively accessible, but unfortunately, it is far more difficult to measure. But whether the reference is to narrow or to broad heritability — indeed, in all technical discussions of the relation between genetic and phenotypic variation — a crucial distinction divides both variants of the term (h and H) from the colloquial meaning: the technical definition (or definitions) is a statistical rather than a causal measure. In other words, it has meaning only in relation to the properties of a population, not to properties either of an individual or of an individual lineage.

In contrast, the ordinary meaning of the term clearly refers to one or more properties of an individual in relation to his or her ancestry, and it is equally clearly aimed at the question of cause, or origin, of a particular trait in the individual: How did Bill or Tom come by his complexion, artistic talent, or fortune? One possible answer is by his inheritance. The trait in question is heritable. But it is not possible to use the noun *heritabil-*

ity, as technically defined, in the same way. As the authors of a 1997 statement issued by the American Society of Human Genetics stress, "heritability is a descriptive statistic of a trait in a particular population, not a trait in an individual" (Sherman et al. 1997, 1266). As a statistical measure, heritability can have meaning only relative to a population of individuals, exposed to a range of environments, with traits that vary from one individual to another. In other words, I can ask if my musical ability (or, in my case, inability) is heritable, but I cannot ask, what is the heritability of my musical ability (or hair color, height, or any other personal trait)?

Although it is enormously important, many people find the distinction between the ordinary and technical meanings of heritability almost impossible to keep in sight, particularly when discussing human behavioral traits. Authors and readers alike routinely slide from one meaning to the other, wreaking havoc on the ways in which legitimate scientific measurements are interpreted. Indeed, I am persuaded that the widespread erasure of this distinction is responsible for a good part of the apparently irrefragable confusion that surrounds discussions of heritability in human populations, and that has surrounded such discussions from their earliest days.

Curiously, however, this problem of unstable interpretation scarcely arises in discussions of heritability in agricultural crops and populations of livestock, and we may ask why that is. The reason, I think, is simple. Agricultural interest in heritability is inherently both pragmatic and statistical: farmers want to know how best to use their limited resources to improve their crops, in order to increase their profits. Or they want to know how to effectively select seeds for better yields. Unlike academic biologists, farmers are seldom concerned about how an individual plant or animal acquired a superior attribute. Their interest focuses on average yields, and their causal inquiry is accordingly confined to populations. Indeed, I suggest that this is why measurements of narrow heritability suffice for them, and why they do not need to engage in the much more difficult task of measuring broad heritability. In any case, the ban on human experimentation precludes measurements of narrow heritability of the kind typically performed by animal breeders, and for that reason alone human behavioral geneticists have been obliged to turn to broad heritability. Furthermore, many harbor the hope that measurements of broad herit-ability — i.e., of the proportion of phenotypic variation that can be at-tributed to the total genetic variation in the population — will provide some causal understanding of how a trait acquired its particular value or form. As Sesardic writes, "Of the two concepts broad heritability is of greater general interest (for it is this notion that reveals how strong the relative influences of genotype and environment are on phenotypic differ-ences)" (1993, 398–99). And because broad heritability is so commonly interpreted just as Sesardic does here, it is this measure on which the remainder of this chapter focuses and to which my remarks about technical heritability will henceforth refer.

Common Pitfalls

The difficulties of both making and interpreting such measurements — and the misuses to which they have so frequently been put — are the subject of a voluminous literature that began even before the introduction of *heritability* as a technical term. Lancelot Hogben's *Nature and Nurture* (1933) may have been the first contribution to this literature; in our own time, we have Richard Lewontin (1974 and 2000), Ned Block (1996), Patrick Bateson (2004), and many others to thank. One point of these critiques is the fact (already mentioned in chapter 2) that calculated ratios of genetic variance to overall variance in phenotype are meaningless in the presence of either statistical interaction (when genetic and environmen-tal variations are correlated) or constitutive interaction (when genetic and environmental effects are intertwined), for under such circumstances, phenotypic variance cannot be partitioned into a genetic component plus an environmental component. This criticism is of particular relevance to human behavioral genetics for the simple reason that such interactions are ubiquitous in the development of human behavior.

Another common theme is the reminder that the significance of such measurements (when they can be meaningfully performed) is entirely dependent on the particular choice of population and the particular range of environments that the members of the population inhabit. For example, a finding of high heritability may allow one to conclude that most of the variation observed in the particular population under study is due to genetic variation, but that conclusion cannot be generalized to other populations with different ranges of genetic properties, or having available different ranges of environmental contexts. Above all, such a finding does not provide an argument for genetic determination. Bluntly put, technical heritability neither depends on, nor implies anything about, the mechanisms of transmission (inheritance) from parent to offspring.

As an example, consider a trait that is known to be biologically inherited (i.e., repeated from generation to generation), such as, e.g., the number of hands an individual has. We would normally say that hand number is a heritable trait. But what is its technical heritability? Answer: zero, or very close to it. And the reason is that, while there is phenotypic variance in the human population (not everyone has two hands), this variance is almost entirely due to accidents, not to genetics. The genetic variance relevant to hand number in the population at large is virtually nil.

Or let us take, one last time, the example of PKU. The expression of this disorder is clearly a heritable phenotypic trait, in the sense that the mutations responsible for that trait (most readily signaled by high levels of phenylalanine) are transmitted from parent to offspring. Yet children raised from birth on a carefully monitored, low-phenylalanine diet will not manifest the high level of phenylalanine that is the primary phenotypic marker of the disease. So let us ask, what is the technical heritability of PKU? The answer depends on the population we choose. If we choose a population of individuals all of whom have a mutation precluding normal metabolism, but only some of whom have been raised on this special diet, and hence only some of whom have the high levels of phenylalanine associated with the disease, the phenotypic variance will be substantial even though there is no genetic variance in the population. Thus, counter to what intuition might lead us to expect, the technical heritability of PKU for this population will be zero.

Furthermore, as many others have already pointed out,[3] there is also the converse problem. Suppose we have a population in which a particular trait — say, IQ — has been shown to have high heritability, and let us, for the moment, take that claim at face value. What, if anything, could we conclude about a genetic basis for IQ? Because we do not know the genes that are supposedly responsible for IQ, it is customary to take a population that is genetically varied in some general sense, perhaps with respect to race or sex. Now we measure the percentage of phenotypic variation due to genetic variation, and we find it is high (say, 50 percent).[4] What this shows is that some genetic variation (e.g., variation in skin color or sex) is correlated with variation in IQ. What does that have to do with the transmission of IQ from generation to generation? Well, perhaps something, but it says nothing about the mechanism of that transmission. For example, skin color (like sex) is genetically transmitted, but there is no reason to believe that either is biologically linked to IQ. The measurement of high heritability might simply reflect the fact that skin color (like sex) provokes certain expectations that are themselves transmitted from generation to generation — not biologically, but culturally, or environmentally. It has been shown that such social expectations can have dramatic effects on performance (i.e., on measures of IQ). So IQ might be passed on from generation to generation, but here the mechanism of transmission is mediated not by causal factors that (like genetic or more generally biological factors) are internally transmitted from parent to child, but rather by causal factors (cultural biases) that are socially handed down through the generations. In other words, we cannot predict the technical heritability of a trait even when it is known to be genetic, and we cannot predict whether a trait is genetic even when it has been shown to have high heritability.

The Central Puzzle

These conclusions are neither original nor, for the most part, controversial; similar arguments have been put forth over and over, often by authors with impeccable credentials. What is most surprising to me about this history of criticism — and what was the original motivation for this book — is its relative ineffectiveness. Notwthstanding the soundness of their logic, these critiques have scarcely made a dent in the general enthusiasm (both popular and professional) for measuring the heritability of human behavior. Furthermore, reports of high heritability continue almost invariably to be read by lay readers as implying that the trait in question is genetic, and in case anyone misses this implication, it is almost always made explicit by the reporter. Indeed, he or she can hardly be expected to grasp the distinction I am making here, when scientific authors and readers so often make the same mistaken inference.

The persistence of such misreadings — even by scientists who at other times seem so clearly aware of the problem — is certainly puzzling, and I believe that it demands an explanation, especially when we consider all the trouble to which such misreadings give rise. Most attempts at explaining this persistence have focused on political interests (especially on vested interests in maintaining the social status quo), and these surely are an important piece of the puzzle. Indeed, the most notorious proponents of the importance of heritability measures (e.g., Jensen in 1969, Cattell in 1972, and Herrnstein and Murray in 1994) drew out the political implications explicitly. But I am convinced that social and political agendas are not all that is at work. At least equally important is the extent to which we are plagued by a linguistic practice that actively elides the distinction between the causal dynamics of individual development and statistical patterns of correlation, and that, by so doing, structures both our basic intuitions and our reasoning. The arguments that have already been put forth by other critics do not need repeating; I do so here merely as backdrop for the analysis (the core of my project) of the workings of this linguistic practice, hoping meanwhile that existing critiques can be both enriched and strengthened by such an understanding.

As I've already indicated, many scientists — especially those concerned with such matters — are well aware of the common pitfalls and, in fact, often explicitly identify them. It is a cardinal sin in this business to draw conclusions about individual inheritance from measures of technical heritability. Yet it is precisely such inferences, drawn by these same scientists, that one frequently finds in both the semipopular and the technical literature. Or perhaps one should not say that these inferences are actively drawn by the authors in question, but rather that they are carried by the language the authors conventionally employ.

For example, in his widely read *The Blank Slate*, Steven Pinker — following an earlier suggestion by Eric Turkheimer and Irving Gottesman (1991)[5] — claims that the "first law" of behavioral genetics is that "all human behavioral traits are heritable" (2002, 373). What does he mean by this? Here is a slightly fuller statement from the same book that might help us understand:

> All five of the major personality dimensions are heritable, with perhaps 40–50 percent of the variation in a typical population tied to differences in their genes. The unfortunate wretch who is introverted, neurotic, narrow, selfish, and undependable is probably that way in part because of his genes, and so, most likely, are the rest of us who have tendencies in any of those directions as compared with our fellows. It's not just unpleasant temperaments that are heritable, but actual behavior with real consequences. Study after study has shown that a willingness to commit antisocial acts . . . is partly heritable (though like all heritable traits it is exercised more in some environments than in others). (ibid., 50)

The word *heritable* here will almost surely be read (and I suspect Pinker intends it to be read) in the usual sense of transmissible, and more specifically, in what has become the usual sense of biologically transmissible — i.e., genetic. But what are the studies underpinning Pinker's colorful claim, studies that have shown personality traits to be heritable? They are not studies in experimental genetics; rather, they are measures of technical heritability.

In fact, the same problem persists in the technical literature of both behavioral genetics and quantitative population genetics. It is quite common to read in published reports of heritability measurements claims, based solely on (technical) heritability measures, such as "behavioral components of Bulimia nervosa (such as self-induced vomiting) are heritable" (Bulik et al. 2003); "verbal and spatial ability are recognized as highly heritable" (Bratko 1996, 621); longevity is "only moderately heritable" (Herskind et al. 1996, 320); and "Hyperactivity in Pre-school Children Is Highly Heritable" (Price et al. 2001). Additional examples abound. Furthermore, in a classic paper by Plomin and his colleagues, such measures are said to identify not only the "most heritable domains of behavior" (Plomin, Owen, McGuffin 1994, 1738), but even "The Genetic Basis of Complex Human Behaviors."

Indeed, even the dictionaries can lead us astray. My earlier listing of technical definitions of heritability shows only some of the definitions I collected — I deliberately omitted those that actively perpetuate the confusion at issue. Here are two I left out. In the eminently respectable compendium, the *Dictionary of Genetics*, published by Oxford University Press, we are told: "in the broad sense, heritability is the degree to which a trait is genetically determined, and it is expressed as the ratio of the total genetic variance to the phenotypic variance (V_g/V_p)."[6] My other omission was one of the alternative definitions given by the *Glossary of Genetics*, a mainstay of genetics dictionaries that has been in print since the early 1950s: "the degree of genetic control of the development of a character." Clearly, in these definitions of heritability, the crucial distinction between individual lineages and populations of lineages, or between the ordinary and technical sense of the word — the very distinction that is required to make sense of its uses — has all but vanished.

On the Instability of Language and the Matter of "Mistakes"

So what is going on here? How could so many eminent scientists possibly be wrong? Or perhaps I should ask, how could so many eminent scientists be *simply* wrong? The answer is that they are not. There is a mistake being made, but it is far from being a simple mistake. Let me try to clarify exactly what I am claiming.

I am not claiming that measures of heritability have nothing to do with heredity, for they certainly do. The problem is to clarify, first, what kind of heredity (genetic, epigenetic, cultural, etc.) one is speaking of and, second, exactly what bearing the statistical measure has on that kind of heredity. Measures of technical heritability are clearly related to the transmission of differences, or variation, in reproducing populations. My concern is with what they are not about. And what they are not about is either the causal basis of individual traits (the problem on which chapter 2 focused) or the direct transmission of individual traits in individual lineages (the problem of this chapter). That is, they are not about the heritability of traits in the ordinary, nontechnical sense of that term. If such a distinction seems difficult to see, if my claim seems counterintuitive, at least part of the reason is that our intuition itself has been trained by the words we use. And given that such mistaken inferences are so difficult both to detect and, once identified, so difficult to keep hold of, I need to try to locate exactly where it is that an error has occurred. This, I fear, not only requires some belaboring of the issue, but it also raises the interesting question of whether or not clarity is even possible. Can a stable distinction between individual and population in fact be maintained in genetics?[7]

Most behavioral geneticists would agree that a mistake has been made when explicit claims about the genetic basis of individual traits are inferred from measures of technical heritability, or at least I hope they would. More commonly, however, such claims are formulated simply in terms of heritable traits, with the implication of a genetic basis left just short of explicit. Large values of heritability are interpreted as showing that a trait is partly heritable, highly heritable, or simply just heritable. Has a mistake been made here, and if so, precisely where? Well, that depends on what one means by *heritable*. If one reads such claims in the ordinary, nontechnical sense of the term, then yes, a mistake has been made. But if one were to confront the authors of such claims (as I have) with this obvious source of confusion, a common response is: "Yes, of course. But that's not the way I use the term; I use it merely to refer to measures of heritability in the population."

In fact, one can point to numerous examples in the literatures of quantitative and behavioral genetics in which measures of heritability are said to reflect "the degree to which a trait is heritable." Even the popular textbook *An Introduction to Genetic Analysis* — the text that provides what is probably the most careful analysis of heritability available — uses the term *heritable* in this way. For example, the authors explain the important difference between *familiality* and *heritability* as follows: "Character states are *familial* if members of the same family have them in common, for whatever reason. They are heritable only if the similarity arises from having genotypes in common" (Griffiths et al. 2008, 652).[8] And for emphasis, a highlighted message reads: "the question 'Is a trait heritable?' is a question about the role that differences in genes play in the phenotypic differences between individuals or groups" (651).

This is clearly a use of *heritable* that is quite different from the ordinary sense of the term, and in principle, it is hard to object: scientists frequently redefine ordinary terms (as of course they did with heritability itself). Here, heritable is not used to mean transmissible; it is no longer a property of individuals or individual lineages, but rather one of populations — or, at the very least, of pairs of individuals. Using the term in this sense, I can no longer say, e.g., that my eye color is heritable. Now, a trait is heritable if and only if its (technical) heritability is greater than zero.[9] It is noteworthy, however, that this shift in meaning is nowhere explicitly signaled.[10] Furthermore, even if the new meaning were to be made explicit, the problem of knowing in which sense the term is being used at any given time would remain, and it is precisely the not knowing that makes the distinction so difficult (if not impossible) to maintain. Finally, even if behavioral or quantitative geneticists were scrupulously careful to avoid such ambiguity, never sliding into the colloquial sense of heritable, as soon as we turn to neighboring literatures in molecular biology and much of medical genetics, the slippage immediately returns, for there we find quite a different convention in the use of the term — still technical, but now clearly in accord with its colloquial meaning.[11] According to Susan Lindquist, for example, entities are heritable if they "span generations and influence a cell's phenotype in a predictable way."[12] Or, in a report recently issued by the National Institutes of Health (2007), we read: "More than 200 heritable disorders of connective tissue (HDCTs) affect the tissues between the cells of your body. The disorders are called 'heritable,' because they are passed on from parent to child."[13] As used here, the notion of a heritable trait has nothing at all to do with statistical studies of genetic and environmental variance in populations — i.e., with measures of technical heritability; rather, the claim that entities (traits, disorders) are heritable is based on studies of lineages.

Finally, even apart from its manifest ambiguity, I'm not exactly sure how to parse the term in a strictly populational sense. For example, what might it mean to say "a trait is statistically heritable," without implying anything about its transmissibility from parent to offspring? Not much, in fact. To make such a sentence meaningful, we would have to recast it to something like "the difference, or variation, between traits is heritable" — i.e., passed on from a population of parents to a population of offspring.[14] In other words, the variation in a population is inherited from the variation in the parental population, and locutions claiming that traits are heritable ought to be read as shorthand for statements claiming that variation in traits is passed on through the generations. Just how that variation is passed on (or inherited), we cannot say. It might involve the direct passing on from parent (or from the union of two parents) to the one or more offspring they produce of some attribute — e.g., an allele — that is associated with a particular variant, or it might require a much more complex (and indirect) pattern of transmission requiring the participation of many individuals (as, e.g., is required for all traits dependent on a social context). On the question of the mechanism of transmission, measures of heritability are simply silent.

What about the various other locutions that are often employed — for instance, what might it mean to say a trait is highly or partly, or only slightly heritable? Are we speaking about something like Darwin's "strength of the hereditary tendency"? If so, in what does this strength inhere? That is, are we speaking of the strength of the hereditary tendency in individuals? In lineages? In populations? Darwin himself invoked this notion only once, and he did so while discussing the operation of artificial selection on variation. He argued that "the strength of the hereditary tendency . . . [coupled with endless variation, renders] the whole organization . . . in some degree plastic" (1869, 62). By "the whole organization," he clearly referred to the breeding pool, and by "the strength of the hereditary tendency," its responsiveness to selection. Darwin of course did not know about DNA, and he had little to say about the actual mechanism of inheritance (i.e., about genetics as we know it); he referred instead to phenotypic preservation and transformation. Nevertheless, there is an interesting resemblance between his formulation and the efforts of today's population geneticists to measure the strength of heritability (in the ordinary sense of the term), solely on the basis of phenotypic measurements. Darwin surely had populations in mind when he invoked the notion of strength, just as today's population geneticists do (at least most of the time) when they speak of traits being relatively heritable. Furthermore, as soon as one attempts to quantify the degree to which a trait is heritable, the shift to populations seems unavoidable, and the slide from notions of the strength of individual heredity to statistical measures of technical heritability almost natural. Indeed, these notions relate so closely to one another that a simple formulation of the relation between technical and colloquial heritability suggests itself—namely, that technical heritability might be understood as the "strength" of colloquial heritability in a population.

Jean Gayon has argued that the notion of a hereditary force as a measurable quantity, the strength of which could be assessed through statistical analysis, was widespread among breeders in the second half of the nineteenth century, and further, that this notion underlay the fundamental divide later to emerge between the biometricians and the Mendelians.[15] In Gayon's view, the emphasis on lineage over populations so characteristic of the Mendelians, and of classical genetics in general, went hand in hand with the view of heredity not as a force, but as a structure; as residing in material entities (or substances) that could be transmitted from one generation to another (Gayon 2000). However, this view of heredity was not subject to quantitative measurement. The obvious exception would appear to be the technical concept of heritability, which, he argues, "corresponds quite well to what the 19th century scientists had in mind when they characterized heredity as a force of varying intensity" (83). But, for many of the same reasons as I have advanced here, Gayon concludes that heritability, "the only magnitude that modern genetics may appear . . . to offer as a 'measurement of heredity' is not a good candidate for this status. True, heritability is a measurable magnitude, but it is incorrect to say that it would measure 'heredity'" (85). Similarly, I argue that it remains unclear what measuring the strength of heredity without recourse to statistics would mean—i.e., what it would mean to measure the strength of inheritance in an individual lineage.

On the Difficulty of Knowing What One Means

In the table below I summarize my own efforts to translate the terms *heritable*, *heredity*, and *the strength of heredity* as they are employed by experimental geneticists, molecular biologists, and quantitative and behavioral geneticists to refer, alternatively, to individuals or lineages and to populations.

Semantics of *Heritable* and Related Terms

	LINEAGE	POPULATION
Trait *x* is heritable	Transmission/inheritance either of a trait or of the underlying causes of that trait between generations Genetic	"The question 'Is a trait heritable?' is a question about the role that differences in genes play in the phenotypic differences between individuals or groups."[1]
Differences (variation) in trait x *are heritable (heritable variation)*	*Mutations are inherited variations (H. J. Muller 1922)*	*Genetic variance is preserved/inherited*
Trait *x* is highly heritable (or *x* is heritable to a high degree)	None	A high proportion of the variation in trait *x* is genetically inherited (*x* is therefore subject to selection); heritability
"strength of heredity"	*None*	*"narrow hereditability"*

[1] Griffiths et al. 2008, 651.

The question is, how is a reader (or, for that matter, an author) to know which meaning is intended in any particular context? Is it even possible to keep the different meanings of *heritable* apart? The problem is that, as the different meanings of the term travel back and forth between different kinds of arguments, different logics, and different disciplines, the ensemble becomes knitted together into a seemingly coherent whole, giving rise to a seemingly coherent argument. But is the argument that emerges really coherent? Does the whole even exist? The usual response of scientists to analyses of ambiguity and polysemy in their language is that, notwithstanding such ambiguity, they know what they mean. But I wonder. When what we say is inherently ambiguous, when the words we use are themselves polysemic — i.e., when they have multiple meanings — is it possible to know what we mean? Or is it not inevitable that some of semantic multiplicity of our terms generates a multiplicity, and hence an inherent indeterminacy, of meaning? Certainly, we generally try to be clear about what we mean, and scientists are especially committed to semantic univocality. And sometimes, perhaps often, they succeed. But my claim is that, when the words they use have multiple meanings, meaning is not so easy to control, either in the minds of their readers or in their own minds. Consciously or not, slippage happens; as a consequence, it is not only easy to mean two — or even three — things at once, it may be unavoidable. What is difficult is meaning only one thing.

For example, I would argue that many behavioral and quantitative geneticists actually mean that a high measure of a trait's heritability, while clearly referring to a population, also has implications for the causal dynamics of individual traits. Precisely because the meaning of the term is uncontrollable, the dual implication of heritability — of transmission from generation to generation both of genes and of genetic variance — is almost impossible to avoid. Moreover, I suggest more generally that much of our interest in heritability — as well as many of the arguments put forth on behalf of the importance of measuring such a quantity — rests deeply and inextricably on the unspoken ambiguity of the terms, and on the slippage that ambiguity invites. It rests, in short, on the informal fallacy that formal logic calls equivocation.[16]

Is There a Solution?

Should we attempt to introduce a different term for the technical quantity we now call heritability? That might help, but I suspect, as I have already suggested, that the ambiguity between the colloquial and technical meanings of heritability has been important to people precisely because it works so well to bolster the basic dichotomy that was introduced at the end of the nineteenth century. I will argue in chapter 4 that a shift in the vocabulary of genetics may be more to the point. If we want to know what makes us what we are, what we need to understand is how these various systems of heredity interact to shape our development, both ontogenetically and phylogenetically. If we want to know what makes us different from one another, then here too the complexities of development must be taken into account. And recent insights about the mechanisms of development suggest that there may be ways to formulate both of these questions in ways other than those that have so far dominated our attention — ways that are likely to be more productive precisely because they are more scientifically accessible.

FOUR What's to Be Done?

How can we extricate ourselves from all these confusions and equivo-cations? Like many others, I would love to see the nature-nurture debate just fade away. But I recognize that that is not likely to happen until we are able to identify the questions in which people are interested, and that at the same time are addressable by scientific research, and separate these questions from the ones that are not answerable. Up to now, I have focused on the slippages and muddles that keep the debate going, but equally crucial to its persistence is the fact that, however confusedly, the debate does seem to capture a number of issues that many people want to know about. It is now time to try to identify these issues, and to reformu-late them in ways that might permit some resolution through the tools of modern genetics.

The stress on modern genetics is crucial. A major part of my argument is that, although the concept of the particulate gene that we inherited from the early days of genetics may have helped establish some of the basic kinds of equivocation we have seen, the new science of genetics coming out of today's research laboratories may point us to a route out. The aim of this final chapter is to sketch some of the ways in which that can happen.

The Fixity and Plasticity of Human Nature

Perhaps first on the list of concerns that get included in the nature-nurture debate is a question that is all but explicit in the words themselves, or at least in the meanings that used to be their primary import, and that are often so understood to this very day. I refer to the understanding of nature as prenatal (inborn, congenital, or innate) and of nurture as postnatal (acquired or learned). Many people are clearly interested in distinguishing one set of influences from the other, especially when the definition of nature as inborn and nurture as what is added after birth is combined with the assumption that what is inborn cannot be significantly altered by nurture. This is the concern that is obvious in discussions posed in terms of innate and acquired characteristics. It comes to the fore in expressions like "he was born that way." Conventionally, such a statement is not meant to be taken literally: it does not mean that the trait in question is fully formed at birth. Rather, it means that nothing happening after birth exerted (or could have exerted) any influence on the formation of that trait. It was fixed by his nature. Well, perhaps not totally fixed, but a reasonable paraphrase might be that a child's postnatal experience would have little effect on the development of that trait. For all practical purposes, it was part of his birthright (or wrong, as the case may be).

Some traits are indeed quite difficult to change after birth, and manifestly so. The basic form of our bodies, for example. Or, barring accident, the number of our fingers, toes, or limbs. Or possibly even the range of our emotional responses. All of these fall into the category of traits pretty much fixed at birth. And the question of what other traits might also be (more or less) fixed at birth would seem to follow naturally. But to ask this is a far cry from attempting to separate the effects of pre- and postnatal experience, or to weigh their relative influence. We might be able to clearly distinguish between various factors introduced at different times (say, genetic endowment, prenatal environment, and postnatal experience), but given how interdependent the effects of each of these is on the others, we cannot separate their respective influences on the final outcome. We might be convinced of the importance of Suzy's genetic heritage to her height, and equally convinced of the importance of how much milk she drank as a child — we might even be able to say how much difference the availability of milk made — but we still cannot answer the question of how much of her height is due to her genes and how much to her milk consumption, or even the question of which of these factors was more important to her growth.

If we are interested in what traits are fixed, a far better way to pose this question (better in the sense of more likely to produce meaningful answers) would be in terms of the presence or absence, or even the degree, of phenotypic plasticity. Put in these terms, this is a subject of enormous scientific interest today. But there is still a problem with posing questions about phenotypic plasticity in terms of pre- and postnatal influences, and it has to do with the relative arbitrariness of birth as a point of demarcation. To be sure, birth brings with it an abrupt change in environment, but that change does not in itself imply a corresponding discontinuity in developmental plasticity.

Research has taught us of the enormous plasticity of many traits (both physiological and behavioral), but in general, plasticity is a function of developmental age. Some traits are relatively difficult to change after birth; other traits become relatively unmalleable only after the first six months of gestation, after infancy, or after adolescence. In this respect, there is in fact nothing special about birth as a cutoff point. Thus to be addressable by scientific research, the question would have to be rephrased along these lines: to what extent is any given trait (more or less) fixed at particular points in development? "More or less" is crucial here, especially for behavioral traits, for we now know that almost nothing in the brain is totally fixed at any point of development, although heroic efforts may be required to produce change. So perhaps we should rephrase the nature-nurture question, and ask, instead, how malleable is a given trait, at a specified developmental age?

This certainly is a legitimate question, and I suggest that it captures the essential features of interest in many of the discussions of the roles of nature and nurture in the formation of individual traits conducted in the English-speaking world between the late sixteenth and late nineteenth centuries. I also suggest that it captures the essential features of interest in contemporary discussions of innate and acquired characters.[1] Finally, it is worth noting that, in this rephrasing, any implication about the causal agency by which a trait comes to be more or less fixed, more or less malleable, has been abandoned. So too has any implication of opposition, or indeed of an either-or relation, between pre- and post-natal influences.

Growing Up and Growing Apart

The confounding of what it is that genes do with the consequences of gene defects, of the difference between the causal impact of genes with the causal impact of mutations (or gene differences), is everywhere. It is part and parcel of the conceptual baggage that has been carried along throughout the history of the gene concept. That it has so deeply penetrated our thinking about nature and nurture is hardly surprising, especially when we consider how deeply wedded the concept of nature has been (at least since Galton) to expectations stemming from the hypothesis of particulate units of inheritance and, from the beginning of the twentieth century, to those stemming from the concept of the gene. But today, that concept is in disarray. With his typical acerbity, Sydney Brenner observes that "old geneticists knew what they were talking about when they used the term 'gene,' but it seems to have become corrupted by modern genomics to mean any piece of expressed sequence" (2000, 2174). Even more dramatically, the molecular biologist Lee Silver suggests that "the notion of the Mendelian gene as a unit of heredity, scientists now realize, is a fiction" (2007). Not all biologists would go so far, but the number of those who would is steadily on the rise.

No one will deny the immense value that the Mendelian gene had for biological research throughout the twentieth century, but increasingly, contemporary scientists recognize that new ways of conceptualizing inheritance are required to make sense of the new data that are pouring in from research laboratories. Recent work on inheritance systems that do not operate through the replication of DNA sequences (epi- or nongenetic inheritance) has added several new dimensions to our understanding of inheritance.[2] But even if we restrict our focus to the transmission of DNA sequences, new research has taught us the importance of a kind of extra-genetic inheritance that comes to the fore only when we have fully recognized how much more DNA contains than protein-coding sequences (the most common association with the term *gene*).

I am referring here to recent discoveries of both the quantity and the biological importance of noncoding DNA that is faithfully transmitted through the generations. We now know, for example, that only 1–2 percent of human DNA is devoted to protein-coding sequences, which leaves the vast majority of DNA unaccounted for. In the past, the noncoding DNA that we knew existed was dismissed as "junk" (leaving unanswered the question of why it persists), but now geneticists are beginning to learn of the enormously complex biological functions that these sequences are involved in. In addition to being transmitted through the generations, it seems that most of the noncoding DNA is transcribed into biologically active RNA sequences that can play crucial roles in individual development. Thus, when medical geneticists succeed in correlating anomalies in the DNA sequence with disease states, the overwhelming likelihood is that these anomalies are not in what most people mean when they speak of genes. This does not mean that these anomalies do not have serious consequences, it only means that we do not yet understand what these consequences are, and to speak of the aberrant sequences as genes serves only to obscure our ignorance.

Today, when biologists refer to the importance of genetics, they are generally referring to DNA rather than genes, and confusion would be greatly diminished if talk of genes were to be replaced by talk of DNA. We may no longer know just what we mean by the word *gene*, but we do know what we mean by the term "DNA." We know how to sequence it, we can study when and where it is expressed (i.e., transcribed), and, as a result, we are beginning to understand its immensely complex role in both development and evolution. Especially important is the understanding we are beginning to acquire of at least some of the roles played by those parts of DNA that do not code for any proteins, yet are nonetheless faithfully transmitted through the generations. We do not know the full function of such extragenetic DNA, but we do know that earlier attempts to dismiss it as junk were mistaken. Daily, we are discovering new and extraordinarily ingenious ways in which noncoding DNA sequences participate in the mammoth projects of regulating the spatially and temporally specific transcription of DNA, the construction and translation of messenger RNA, and the positioning, conformation, and activity of proteins. Early concepts of the gene were predicated on the assumption of a relatively simple transformation from genotype to phenotype, but now we are beginning to understand just how enormously complex that process is, and how demanding it is of ongoing fine-tuning and regulation.

Such findings not only require us to rethink basic assumptions in biology, they also create the opportunity for such reconceptualizations. In particular, I suggest, they can help us extricate ourselves from at least some of the confusions about nature and nurture that became so endemic in the last century. They invite us to try to understand how traits are formed by focusing not on the putative causal powers of such conceptual entities as genes, or even on the concrete and incontrovertibly real sequences of the DNA we inherit, but on the interactive dynamics of the extraordinarily resourceful molecular systems that evolution has bequeathed to us.

To be sure, none of the complexity of these dynamics calls into question the fact that simple changes in the base sequence of DNA can have major effects on the formation of traits, whether or not these mutations are located in protein-coding sequences. Furthermore, many diseases are undoubtedly due to errors in one or more protein-coding sequences. My argument does not challenge the claim (returning once again to the familiar example of PKU) that a mutation in, say, the sequence coding for the enzyme phenylalanine hydroxylase can have disastrous consequences for human development. Nor does it call into question the role of protein errors in any of the other single-gene disorders that have been so widely publicized despite their relative rarity—such as achondroplasia, cystic fibrosis, hemophilia, Huntington's disease, sickle cell anemia, and Tay-Sachs disease). If such conditions have received a disproportionate amount of attention it is, at least in part, because the role of mutations in protein-coding regions were what scientists have been able to study. Only recently has the importance of mutations along the vast stretches of DNA that do not code for proteins begun to be appreciated, and has become accessible to investigation. Of course, any disease that is triggered by a mutation in DNA might be said to be genetic, but what cannot be taken for granted is that the trigger is in a gene.

Furthermore, even when genetic, a disease's onset and severity might nonetheless be subject to the influence of all sorts of extragenetic conditions, like diet, drugs, or other forms of therapy. Huntington's disease is of particular interest to many scientists because of its high penetrance — i.e., the overwhelming likelihood that the disease will manifest itself over the course of a lifetime — but even here, the prospect of stem-cell therapy holds out hope for many patients. Indeed, as Philip Kitcher (1996) has asked, should we not view the therapeutic injection of normal cells that have not been inherited, that are not inborn to the affected individual, as itself an "environmental" effect?

In addition, I need to emphasize that my argument in no way undermines the importance of DNA's role as a molecular resource for the operation of biological systems; that importance, we can agree, is not likely to be supplanted. Even if it may no longer make sense to think of DNA as the "master molecule" of biology — the molecule that encodes and orchestrates the dance of life — DNA must still be recognized as one of the most critical components (if not the most critical) of the cell. Indeed, as I have suggested earlier, DNA has turned out to be a far more interesting molecule, providing a far richer reservoir of potential applications, than anyone had imagined. It cannot be divided into fixed, prescribed units called genes, nor can it animate the cell, but it does constitute an astonishingly fertile resource both for the construction of the many kinds of actors that do animate the cell, and for the systems of regulation the cell employs for its maintenance and reproduction. In all of this, the locus of action and agency may have shifted, but the importance of DNA has not. In retrospect, we can recognize that the early mantra of "DNA makes RNA, RNA makes protein, and proteins make us" was not only too simple, but also did not begin to do justice to the wealth of possibilities this molecule actually offers.

Heritability Redux

The problem I focused on in chapter 3 is the chronic slippage between the two meanings — ordinary and technical — of *heritability*. Could not the difficulty be avoided by introducing a different term for the technical quantity that behavioral and quantitative geneticists seek to measure? That might help, but as I have already suggested, the ambiguity of the term has been important to people precisely because, on the one hand, it promises to teach us something useful about inheritance, and on the other hand, it works so well to bolster the basic dichotomy that was introduced at the end of the nineteenth century between nature and nurture. A shift in the vocabulary of genetics may be more to the point. I have already suggested that we accept (perhaps even embrace) the enormous difficulty that contemporary researchers have in agreeing on a single, shared meaning to the word *gene,* and move ahead; that we shift our focus instead to the more directly functional entities that contemporary research has begun to identify. Here I would like to suggest that we also embrace the ambiguity of the term *heritable,* and even extend it. Doing so would not solve the problem of the perennial slide between the technical and colloquial uses of the term, but I believe it would undercut the basic dichotomy that slippage serves.

In other words, let us acknowledge that, with the exception of such accidental losses (like loss of fingers or hands in industrial accidents) that are clearly not transmitted, almost all human traits *are* transmitted from one generation to the next; but at the same time, let us also accept the fact that the mechanisms of transmission are very varied. They may be genetic, epigenetic, cultural, or even linguistic. With this extension, we might even embrace Pinker's first law: "All [or almost all] human behavioral traits are heritable." Poverty is heritable, and perhaps even intelligence (certainly levels of education are); abusive behavior is also heritable, as are, of course, racism and sexism. The problem is that, having agreed that all (or almost all) human traits are heritable, what have we said? Not much.

Which brings me back to the questions of what we want to know, and of how we can legitimately use our scientific resources to give us meaningful answers. I have argued that, to the extent that our interest is in what makes us what we are, it is useless to proceed by trying to separate nature from nurture and looking at how they interact. The causal effects of nature and nurture on development are simply not separable. But some things are separable, and for these, scientific analysis is supremely useful. For example, DNA, RNA, and proteins are molecules that can be studied in isolation. Such studies may not inform us about the many sorts of interactions the molecules are capable of, yet some of these interactions can themselves be studied in isolation, pair by pair. Some of the entities upon which the development of traits depends (like the sequence of nucleotides that codes for a particular protein, together with the molecular machinery required for translation) come into being only as a consequence of these interactions, but once they have emerged, they too can often be studied separately. Inheritance depends on the transmission through the generations of sequences of nucleotides that code for particular proteins, along with sequences that regulate the production of these proteins. Similarly, chromatin marks are sometimes present, other times not, and these too can sometimes be passed on through the generations; the particular causal impact made by the presence of each such marker can be tracked as well. Also, differences in diet, habitat, and cultural norms can be distinguished from one another and the effect of such differences on developing traits can be separately studied. All such effects — whether they derive from nucleotide sequences, interactive protein-DNA networks, inherited chromatin markers, diet, habitat, or cultural norms — are, when they are inheritable, parts of systems of heredity that can be studied independently of one another; furthermore, it is also possible, indeed necessary, to study the effects of the different systems of heredity upon one another.

This is especially important for understanding the evolution of traits, in humans as in other organisms. In that context, it has traditionally been seen as crucial to distinguish between the mechanisms of inter-generational inheritance and the factors that contribute to individual development, whether or not these factors are inherited. More recently, however, the fact that the mechanisms of development are themselves products of evolution has become increasingly apparent, and the boundary between evolution and development has blurred significantly. But for studying the character of developmental mechanisms in themselves, for understanding what makes us what we are in each generation, the question of whether or not the factors contributing to individual development arc hereditary is of less importance, as is the question of just how inheritance occurs (i.e., whether it is genetic or extragenetic). What is of primary importance is learning what the relevant factors are, and how they interact.

Recent successes of molecular biology have shown these processes to be far more complex than we once expected them to be, but the same successes have also produced immense advances in the techniques that can be used to investigate the causal dynamics of development. In fact, the most promising, exciting, and pioneering research in developmental biology now focuses on the challenge of understanding these dynamics. Insofar as our ability to intervene in them depends on our understanding, it should be obvious that such work has immense practical significance. And insofar as our social interest is in optimizing the development of individual human potential, it should be equally obvious that this is where our research dollars should go.

As to the question of what makes some of us different from others, here too the complexities of contemporary genetics turn out to be crucial, for they have clearly shown us just how difficult it is to meet the conditions required to make that question answerable by the statistical analyses required for measures of heritability. James Tabery (2007 and 2008) emphasizes the role that disagreements about the magnitude of the associated difficulties played in the historic debate between Fisher and Hogben on the utility of such measures — in particular, they disagreed on the extent to which the effect of changes in the genotype depends on the particular environment, and hence, on whether gene-environment interactions have a substantial effect on phenotypic variance. Tabery's work reveals the decisive importance of the practical context in which such questions are posed. For many agricultural purposes, such effects may in fact be quite small, especially over the range of environmental variables under consideration. In human traits, however, this is often not the case. Take the studies of race and IQ, surely the most notorious example at hand. Here, the presence or absence of racism turns out to be a crucial feature of the environment, for the effect of racial difference on intellectual performance is strongly dependent on whether an individual grows up in a racist or a nonracist environment.

The question lingers, however, of how we are to respond to those for whom, despite all these difficulties, questions of difference still remain pressing. That, I fear, depends on why such questions seem so pressing — i.e., on the social and economic issues that would be at stake if we could meaningfully answer these questions, and on the particular political, economic, and social values we hold dear. In other words, it depends on politics.

The major practical interest driving the search for the relative importance of different causal factors in producing a given phenomenon is to be found in the wish to effect change in that phenomenon. Farmers seek estimates of heritability because they want to improve the yield of their crops; breeders, because they want to increase the production of meat, milk, eggs, etc. And even if we have ostensibly disavowed the aim of breeding better (more productive?) humans, we nonetheless retain the aim of maximizing the well-being of individual beings. Thus, one of the main reasons we continue to ask what, for instance, makes Lisa shyer (or more cheerful, a better athlete, or more mathematically gifted) than Amy is that we want to do what we can to help our children maximize what we regard as their potential well-being. We seek to improve our parenting skills, the educational opportunities we make available, the quality of medical attention on offer — all in the hope that these improvements will enhance the development of traits we value. Gene therapy has not yet proven itself, but probably this too will be an option one day. Of course, all of these interventions depend on the availability of resources (be they psychological, technical, or financial), and their allocation requires choices that are inescapably political. Financial resources are limited, and, like the farmer, we as a society would also like to spend our resources effectively. But unlike the farmer, how we choose to distribute these resources depends on our social and political values. Also unlike the farmer, our choices are collective rather than individual. Furthermore, our goals tend to be more focused on individual traits than are those of the farmer. The obvious question is, how can scientific research help us? To the extent that the choices we make depend on the social, political, and even moral values we hold, more — or better — scientific data will not be of much use. But on some issues, scientific research can clearly help us set priorities. In particular, it can (at least in many cases) shed light on how effective any proposed intervention is likely to be. The question then becomes, what kind of research can give us this kind of information?

Because I believe that, even without the many problems I have discussed in this book, heritability measures are an extremely indirect way to go about providing such information, my preference would be to reformulate the question altogether, changing it to one that directly focuses on what I think people want most immediately to know, and also one that the realities of biological development permit scientists to actually answer. Let us ask not how much of any given difference between groups is due to genetics and how much to environment, but rather how malleable individual human development is, and at what developmental age. As I have already said, there is no reason to privilege birth as a cutoff point — development is lifelong, and so is its plasticity. We may not share the interests of breeders in artificial selection, but as both scientists and citizens, we are surely committed to trying to maximize the development of individual human potential. And for this, we need a better understanding of what resources can contribute to such development, and of how they can best be deployed. What kinds of research can provide us with such information? I would put my money on the new studies of phenotypic plasticity that we are beginning to see not only in developmental biology, but also in neuroscience, physiology, and ecology. There is no shortage of scientific work that can productively inform us about the things we want to know, but we do need to pose our questions in ways that researchers can answer.

Introduction

1 For a particularly valuable review of such alternative "systems of inheritance," see Jablonka and Lamb, 2005.
2 Especially with recent discoveries of the many different roles played by non-coding DNA in a cell, the definitions of *gene* have proliferated wildly (see, e.g., Keller and Harel 2007).

ONE Nature and Nurture as Alternatives

1 For the Confucian literature, see, e.g., Graham 1986; Ames 1991; Ng 1994; Bloom 1994 and 1997; and Xunwu 2000. For a more general and comparative discussion of ancient Greek and Chinese views of the matter, see Lloyd 2007. Particularly striking is Graham's claim that "analogies for human nature in Mencius are always dynamic, trees growing on a denuded mountain, ripening grain, water finding its natural channels" (Graham 1986, 43). Lloyd's discussion is framed in terms of nature and culture, and his conclusion is to the point: "The nature/culture division turns out to be a far more problematic framework than is sometimes realized for the discussion of the views in question, namely about what there is and about the different categories into which that falls" (2007, 132–33).
2 I take the phrase "nurture nature" from Ng's translation of *yang-hsing* in his discussion of the concepts of nature and nurture in Confucian philosophy (Ng 1994); see also Chen, who argues that the notion of culture as "the art of cultivating nature" (2000, 109) lies at the core of Confucianism.
3 John Locke, March 7, 1692 [i.e., 1692/93], "Some Thoughts Concerning Education," http://www.fordham.edu/halsall/mod/1692locke-education.html.

4 As Paul and Day (2007) make clear, Mill does explicitly disjoin the roles of nature and "habit, custom, and education," or of the innate and the acquired, in the development of differences between individuals and groups, and comments on their relative importance can be found scattered throughout his voluminous writings. Some of his remarks might also, if taken out of context, even appear contradictory. Thus, e.g., in "The Utility of Religion," he argues that "the power of education is almost boundless" (quoted in Paul and Day, 2008), while in *The Subjection of Women*, he admits that "it would of course be extreme folly to suppose that these differences of feeling and inclination only exist because women are brought up differently from men, and that there would not be differences of taste under any imaginable circumstances" (ibid.). I suggest, however, that much of this apparent contradiction is resolved when we separate claims about the kinds of differences that changes in environment can make from claims comparing the relative power of the innate and the acquired, or of nature and nurture, in the development of individual traits. (See chapter 2 for further discussion of this last distinction.)

5 This last sentence was not in the original edition of *Descent* (1871) but was added in the second edition (1879), perhaps reflecting a hardening of Darwin's position over the intervening years.

6 There may, however, have been precursors. For example, Groff and Mc-Rae (1998) have suggested that Heldris de Cornuälle's *Roman de Silence*, a thirteenth-century French romance chronicling the adventures of Silence, a girl turned boy (Silentius) in order to inherit a kingdom, provides a far earlier example of such opposition. The romance does indeed personify nature and nurture locked in struggle: "Nature, leave my nursling alone, or I will put a curse on you! I have completely dis-natured her" (quoted in Groff and McRae 1998, 2593–95). Silence prevails as Silentius, yet her/his ultimate triumph is to be restored to her real gender (and status) when, at the end, she is crowned queen.

7 To my knowledge, the only time Mill uses the word *heredity* is in 1863, in a letter to Alexander Bain, referring to Herbert Spencer's enthusiasm for "the last new theory" (1863, 901).

8 Darwin's thinking about heredity was surely influenced by the development of a concept of hereditary disease in the first half of the nineteenth century. But the notion that normal attributes could be accounted for by physiological heredity was still relatively new in his time. See Beltrán 1994, 2004, and 2007 and Waller 2001 and 2002a and b for excellent discussions of the complex history of the emergence of ideas of heredity in the nineteenth century.

9   Beltrán's argument is compelling. He writes: "The change from adjective to noun points to a change from analogy (or metaphor) to a direct ontological commitment to the reference of the concept . . . [—a shift] to 'heredity' as an explanatory biological concept that implies a particular kind of independent causation" (1994, 214).

10 Over thirty years ago, in his work on nineteenth-century disease and social thought, the historian of medicine Charles Rosenberg made a similar argument, although he dated this change a bit later. He wrote: "Absolute distinction between the innate and the acquired was a concept so novel, so contrary to traditional common sense, that it was not generally assimilated until the second decade of the twentieth century" (1974, 223).

11 Spencer's "physiological units," introduced in *The Principles of Biology* (1864), four years before Darwin's (independent) discussion of "gemmules," are sometimes mentioned as an alternative beginning of particulate inheritance, but in my reading, the point of Spencer's units is entirely different: they provide the basis not so much of a theory of inheritance as of a theory of development, a process that Spencer understood more by analogy with crystallization than as a process of direct determination by heritable units. In any case, Darwin's theory had a more significant impact by far. Mendel must naturally also be mentioned, but his influence was not felt until after 1900.

12 The germ, Darwin argued, "is crowded with invisible characters, proper to both sexes, to both the right and left side of the body, and to a long line of male and female ancestors separated by hundreds or even thousands of generations from the present time: and these characters, like those written on paper with invisible ink, lie ready to be evolved whenever the organisation is disturbed by certain known or unknown conditions" ([1874] 1905 2:26).

13 Darwin speculated that the organism generates its kind not as a whole, but through the production of "minute granules" that were "thrown off" by parts of the organism, circulated through the body, and ultimately collected in the germ cells, from where they are passed on to the next generation. Put in these terms, it made sense to suggest that these particles (or gemmules) could be modified by the changes in the body that had been induced by experience.

14 This reconception corresponds to what, following Ernst Mayr, historians and philosophers of biology generally refer to as a shift from soft to hard inheritance, commonly attributed to August Weismann.

15 *Distinction* is Galton's own term. But if we understand by that word "the action of distinguishing or discriminating; the perceiving, noting, or making a difference between things; discrimination" (according to the online edition of

the *Oxford English Dictionary*), the particular kind of distinction that Galton has in mind ought, given his implication of separability (and hence mutual exclusivity), more accurately be referred to as a *disjunction*.

16 I should make clear that in my discussion of "particulate" inheritance, there is no implication of the transmission of "particulate" traits (for instance, there may, as both Darwin and Galton assumed, be many particles that combine to form each trait), and hence there is no implication of an opposition to what is referred to as "blending inheritance." As I am using the term, *particulate inheritance* refers merely to the transmission of discrete material units assumed to provide the causal basis of the development of traits.

17 Galton's suggestion here is reminiscent of Spencer's (1864) model of development as a kind of crystallization.

18 Galton refers explicitly to advances in physical science to support his conviction of the need of a particulate formulation (1876b, 332). See Stanford (2006) for an especially insightful analysis of Galton's particulate views.

19 My argument is not meant to contradict, but rather to complement, Waller's claim for the importance of Galton's methodology to his thinking about heredity (see Waller 2002b).

20 The issue of the invariance of the causal elements is certainly crucial to the debate over the question of the inheritance of acquired characters that later loomed so large, and here too Galton's views departed from those of his cousin. Indeed, Galton's insistence on the invariance of the germ material might be said to have anticipated August Weismann's arguments. But for the present discussion, the more relevant point is that such an assumption greatly supported (as it was supported by) Galton's commitment to the separability of hereditary factors from environmental influences.

21 See Waller 2001; other sources include Rosenberg 1974; Olby 1985; and Bowler 1989.

22 It is likely that Galton's general anxiety on this score was affected by the failure of his own marriage to produce offspring, as well as by the sterility of the marriages of two of his brothers.

23 The shift in terminology from "nature versus nurture" to "heredity versus environment" is of interest in itself and certainly needs further investigation. It is clear, however, that that shift was substantially encouraged by the resonances of the latter phrase with a broad range of discussions of the time, in many different countries (e.g., England, France, Germany, and the United States) in developmental psychology on the one hand, and about race and milieu or environment in anthropology and anthropometry on the other hand. By the end of the nineteenth century, virtually all of these discussions

had come to be framed in terms of the relative importance of heredity and environment, and references to Galton's methods as a way of resolving disputes on the matter were common. As Franz Boas explained: "If it can be shown that there is a strong tendency on the part of the offspring to resemble the parent, we must assume that the effect of heredity is stronger than that of environment. The method of this investigation has been developed by Francis Galton and Karl Pearson . . . and we look forward to a definite solution of the problem of the effect of heredity and of environment through the application of this method" (1899, 101).

24 Quoted in Pinker 2004, 8.

TWO Changing the Question

1 · This is more an empirical (and practical) issue than a logical one, for it depends on whether gene-environment interactions make a significant difference over the range of environmental parameters deemed relevant. I will discuss this issue at greater length in chapters 3 and 4, but for now, it is worth mentioning that the question of the magnitude of this statistical effect has historically been deeply intertwined with the interdependence of genes and environment in development. For an illuminating discussion of just how this intertwining played out in the debates between Fisher and Hogben, see Tabery 2007 and 2008.

2 Such expectations may of course continue to lurk in the background; indeed, the suspicion that they do accounts for at least some of the political charge attached to these questions.

3 http://en.wikipedia.org/wiki/Nature_versus_nurture, accessed February 8, 2009.

4 To be fair to Woodward, it should be pointed out that he is concerned with only a very limited meaning of "x causes y," namely, that "x is causally relevant to y," and if this is all one wants to infer, I have no argument with the replacement of such a claim by the claim that "a difference in x causes a difference in y" (Woodward 2003, 40). The problem arises when one takes "x causes y" in the stronger sense required for a measure of the relative importance of x as one of two or more causal factors collectively responsible for y.

5 In an earlier work (Keller 2000, chapter 4), I refer to the tacit reasoning behind such inferences as a "logic of subtraction"—likening such an effort to the attempt to infer the operation of a car from all the ways in which the car can be caused to malfunction.

6 See chapter 3 for a discussion of the terms *broad* and *narrow heritability*.

7 Judson proposes as a solution to this problem that we "revive and put into public use the term 'allele.' Thus, 'the gene for breast cancer' is rather the allele, the gene defect — one of several — that increases the odds that a woman will get breast cancer" (2001, 769). It seems to me, however, that the equation of *allele* with *gene defect* risks perpetuating precisely the same confusion: it is not the allele itself that is responsible for the phenotypic difference, but the difference between alleles.

8 Moss writes that "Gene-P" is a "phenotype predictor" and cannot be defined by its nucleic acid sequence. But in fact, "Gene-P" is neither a gene nor a predictor of phenotype: it is a phenotype difference predictor. Indeed, as he himself acknowledges, the reason that "Gene-P" cannot be defined by a specific sequence is that "invariably there are many ways to lack or deviate from a norm" (Moss 2003, 60).

9 Of course, the concept of normal is itself fraught with difficulty, subject to its own ambiguities that primarily have to do with persistent confusion between properties of individuals and those of populations. But most commonly, it too is understood as a relational property, pertaining not to comparison between individuals but to the statistical norm of a population. For instance, Durkheim wrote: "The state of health, insofar as it can be defined, never conforms exactly to that of an individual subject, but can only be established in relation to the most common circumstances" [L'état de santé, tel qu'elle le peut définir, ne saurait convenir exactement à aucun sujet individuel, puisqu'il ne peut être établi que par rapport aux circonstances les plus communes ([1894], 62; my translation)]. See also Hacking 1990, 160–64.

10 The full quotation in the original reads as follows: "On voit qu'un fait ne peut être qualifié de pathologique que par rapport à une espèce donnée. Les conditions de la santé et de la maladie ne peuvent être définies in abstracto et d'une manière absolue. La règle n'est pas contestée en biologie; il n'est jamais venu à l'esprit de personne que ce qui est normal pour un mollusque le soit aussi pour un vertébré. Chaque espèce a sa santé, parce qu'elle a son type moyen qui lui est propre, et la santé des espèces les plus basses n'est pas moindre que celle des plus élevées . . . Le type de la santé se confond avec celui de l'espèce. On ne peut même pas, sans contradiction, concevoir une espèce qui, par elle-même et en vertu de sa constitution fondamentale, serait irrémédiablement malade. Elle est la norme par excellence et, par suite, ne saurait rien contenir d'anormal" (http://classiques.uqac.ca/classiques/Durkheim_emile/).

11 E.g., in *Essai sur Quelques Problèmes Concernant le Normal et le Pathologique* (Essay on Several Problems Concerning the Normal and the Pathological; 1943), Canguilhem writes: "We think with Goldstein that the norm concern-

ing pathology is above all an individual norm" (72; my translation). See also his essay, "Le Concept et la Vie" (1966).

12 For an extensive discussion of the relation between the languages of medical genetics and classical genetics, see Childs 1999.

13 This is not to suggest that maintaining such a diet is easy, or that the almost inevitable relapses are not without dire risks. Probably the best discussion of the history and politics of PKU, as well as of risks associated with its treatment, is to be found in the work of Diane Paul (see, e.g., Paul 1998b and 2000, and Paul and Edelson 1997).

THREE From Individuals to Populations

1 In particular, I might mention an article by Scott Stoltenberg (1997) that I came upon after writing this chapter. Stoltenberg's work covers some of the same ground as I do, and is especially valuable for the many examples it offers of the different uses of the terms *heritable* and *heritability*.

2 Variance is the statistical measure of the average deviation of a variable in a distribution from its mean value.

3 See, e.g., Jencks 1992; Block 1996; Sober 2001.

4 Measuring this ratio is hardly a trivial feat. One way that is often used in work with experimental organisms is to assign a range of values to a particular phenotypic trait, breed a population of pure lines (homozygous genotypes), cross them (in pairs) to obtain a set of heterozygous genotypes, and measure the phenotypic variance of the trait within a population of identical heterozygous genotypes growing in different environments. Because there is no genetic variation in these populations, the phenotypic variance is assumed to be the same as the environmental variance. This value (averaged over the different heterozygote populations) can then be subtracted from the phenotypic variance of the original population (of pure lines) to obtain an estimate of the genotypic variance in that population. Alternatively (and virtually the only way available in studies of human populations), estimates are obtained by comparing the correlations of a trait in monozygous twins raised by their biological parents with the correlations observed in dizygotic twins, also raised by their biological parents. Because monozygotic twins share (almost) all of their genes, and dizygotic twins share only half, the difference between the correlations is presumed to give a measure of heritability. But to the extent that behavioral responses to children is determined by cultural expectations associated with such characteristics as skin color, eye color, and height, monozygotic twins also share more of their environment than do dizygotic

twins. Indeed, it is because of this weakness of twin studies that many quantitative geneticists agree with Richard Mayeux that "heritability estimates do not effectively separate shared genetic from shared environmental influences and cannot effectively apportion the degree of gene-environment interaction" (2005, 1405).

5 Ironically, Turkheimer and Gottesman originally formulated their "first law" as the claim that heritability (in the technical sense) is zero, precisely because (just as with the number of fingers on our hands) there is so little variation in the human population. Indeed, as Turkheimer later elaborated, he saw the law as an attempt to end useless debates about nature and nurture based on measures of heritability, suggesting "that all sides of the issue should stipulate the first law of behavioral genetics and refrain from further discussion of whether or not the heritability of anything is equal to zero. As a reader of behavioral genetics, keep a pencil by your side and lightly excise everything that either asserts nonzero heritability or attempts to explain it away: Much space could be saved in our journals (even those containing sophisticated multivariate genetics or well-informed opposition) if this recommendation were put into effect" (1998, 789). Pinker, in contrast, reads the same law as resulting from the measures of heritability that Turkheimer wishes to excise.

6 This definition may well have been taken from the influential textbook *Introduction to Quantitative Genetics* by D. S. Falconer, where the author writes: "The ratio V_G/V_P expresses the extent to which individuals' phenotypes are determined by their genotypes. This is called *heritability in the broad sense*, or the *degree of genetic determination*" (1981, 113; italics in original). I thank Diane Paul for calling this example to my attention.

7 Of course, the relation between individual organisms and population is enormously complicated by sexual reproduction (see Keller 1987), but I am ignoring those difficulties here in order to focus on the distinction between the inheritance of traits in "individual lineages" (i.e., transmission between parents and offspring) and the relation of the statistical distribution of traits between one generation and the next.

8 This distinction is intended to avoid the confusion between shared environmental and shared genetic influences referred to in note 3.

9 When used in this sense, heritable can no longer be read as inheritable or inherited. This was Gar Allen's point when he wrote: "it is important to point out that the term *heritable*, in the statistical sense, is not synonymous with *inherited*, in the biological sense" (1994, 183).

10 Allen's use of heritable, quoted in the previous note, might be regarded as a

rare exception. The closest I have been able to find for an explicit definition of this alternative use of the term *heritable* is in Griffiths et al.: "the question of whether a trait is heritable is a question about the role that differences in genes play in the phenotypic differences between individuals or groups" (2008).

11 Indeed, as early as 1902, the biometrician Udny Yule complained about the different meanings biologists attributed to the term *heredity*. Ironically, however, he claimed for statisticians the use of the term in the context of what he called "*individual* heredity." He wrote:

> "Heredity" is . . . most usually defined by biologists as "like begets like." In this sense it denotes *inter alia* the phenomenon of the constancy of specific or racial types and of sexual characters; a character may be said to be *inherited* when it always, in one generation after another, is one of the characters of the species, of the race, or of the one sex of the race, as distinct from the other. The species, race or sex, so to speak, "begets its like" as a whole. But then a further question remains; even if the type of the race is constant, do *individual* types within the race beget their like? In so far as any *individual* diverges in character from the mean of the race do his off-spring tend to diverge in the same direction, or not? It is to this question that statisticians have confined themselves, and they speak of a character being "inherited" or not according as the answer to the question is yes or no—they deal solely with what we may term "*individual* heredity." (1902, 195–96)

In fact, however, Yule's reference to "*individual* heredity" was clearly to the divergence of the individual from the norm, and the reference to individual lineages simply does not appear. His concern, like that of Ernst Mayr, was with the distinction between typological and population descriptions. And like Mayr, Yule saw as the distinguishing feature of population analysis its stress on the features of individuals. By contrast, I am focusing on the distinction between populations and lineages of individuals, putting to the side here the literal meaninglessness of the term *lineage* for sexually reproducing individuals (see Keller 1987).

12 "Researchers Show That Proteins Can Transmit Heritable Traits," *HHMI News*, January 27, 2000 (online).

13 National Institute of Arthritis and Musculoskeletal and Skin Diseases, National Institutes of Health, "What Are Heritable Disorders of Connective Tissue?" (2007), http://www.niams.nih.gov (accessed November 15, 2008).

14 However, the notion of heritable variation has also lent itself to differing interpretations. Although it is now often used as a synonym for heritability, to H. J. Muller (1922), e.g., heritable variation referred to mutations that were passed on from generation to generation. What he saw as absolutely central to genes was their "most remarkable property of heritable variation or "mutability" (32), or the transmissibility of mutant forms. He wrote: "inheritance by itself leads to no change, and variation leads to no permanent change, unless the variations themselves are heritable" (35).

15 I accept the general form of Gayon's argument, but I wonder about the implication (perhaps unintended) that we should think of Darwin as exclusively in the latter camp. For many thinkers of his time, the notion of a hereditary force went hand in hand with the assumption that the hereditary force grows stronger over time, and, as Gayon tells us, Darwin was sharply critical of this particular assumption. But the association of heredity with a force does not depend on this assumption. Viewed as a response to selection (as breeders themselves came to do, and indeed as Lush did when he first introduced the new sense of the term *heritability*), Darwin's reference to "the strength of the hereditary tendency" might be said to provide us with a bridge between nineteenth- and (at least some) twentieth-century notions that continue to have great conceptual appeal.

16 Equivocation is defined in the online *Dictionary of Philosophy* (http://www.ditext.com/runes/e.html (accessed August 20, 2009) as "any fallacy arising from ambiguity of a word, or of a phrase playing the role of a single word in the reasoning in question, the word or phrase being used at different places with different meanings and an inference drawn which is formally correct if the word or phrase is treated as being the same word or phrase throughout." Another online philosophy dictionary (http://www.philosophypages.com/dy/e5.htm) illustrates the meaning of equivocation with an instructive example: "Odd things arouse human suspicion. But seventeen is an odd number. Therefore, seventeen arouses human suspicion."

FOUR What's to Be Done?

1 But notice that focusing on plasticity, or malleability, has shifted our attention away from questions about trait development per se to the development of differences in trait development. That is, we are no longer seeking to parse the causal factors of development per se, but rather, we are now looking to sort out different causal influences on developmental differences. To put it in other

terms, we are asking about the effect of changes in the conditions of development on the form of a trait. Such a question might be pursued either through direct experimental intervention (difficult for humans) or through studies of populations of organisms developing under various conditions.

2 See, e.g., Jablonka and Lamb 1995 and 2005.

Allen, Gar. 1994. "The Genetic Fix: The Social Origins of Genetic Determinism."
    In E. Tobach and B. Rosoff (eds.), *Challenging Racism and Sexism: Alternatives
    to Genetic Explanations*, 163–87. Genes and Gender 7. New York: Feminist
    Press.
Ames, Roger T. 1991. "The Mencian Conception of Ren Xing: Does It Mean
    'Human Nature'?" In Henry Rosemont Jr. (ed.), *Chinese Texts and Philosophical
    Contexts: Essays Dedicated to Angus C. Graham*, 143–75. LaSalle, Ill.: Open
    Court.
Bateson, Patrick. 2004. "The Origins of Human Differences." *Daedalus* 133,
    36–46.
Bell, A. Earl. 1977. "Heritability in Retrospect." *Journal of Heredity* 68,
    297–300.
Bickel, H., et al. 1954. "The Influence of Phenylalanine Intake on the Chemistry
    and Behavior of a Phenylketonuric Child." *Acta Paediatrica* 43, 64–67.
Block, Ned. 1996. "How Heritability Misleads about Race." *Boston Review* 20
    (no. 6), 30–35.
Bloom, Irene. 1994. "Mencian Arguments on Human Nature (Jen-Hsing)."
    Philosophy East and West 44 (1), 19–53.
———. 1997. "Human Nature and Biological Nature in Mencius." "Human
    'Nature' in Chinese Philosophy: A Panel of the 1995 Annual Meeting of the
    Association for Asian Studies," special issue, *Philosophy East and West* 47
    (1), 21–32.
Boas, Franz. 1899. "Some Recent Criticisms of Physical Anthropology." *American
    Anthropologist*, n.s., 1 (1), 98–106.

Bowler, Peter. 1989. *The Mendelian Revolution: The Emergence of Hereditarian Concepts in Modern Science and Society*. Baltimore, Md.: Johns Hopkins University Press.

Bratko, Denis. 1986. "Twin Study of Verbal and Spatial Abilities." *Personality and Individual Differences* 21 (4) 621–24.

Brenner, Sydney. 2000. "The End of the Beginning." *Science* 287 (5461), 2173–74.

Bulik, Cynthia M., et al. 2003. "The Relation between Eating Disorders and Components of Perfectionism." *American Journal of Psychiatry* 160, 366–68.

Canguilhem, Georges. 1943. *Essai sur Quelques Problèmes Concernant le Normal et le Pathologique*. Publications de la Faculté des Lettres de l'Université de Strasbourg. Clermont-Ferrand, France: La Montagne.

———. 1966. "Le Concept et la Vie." *Etudes d'Histoire et de Philosophie des Sciences*. Paris: Vrin.

Childs, Barton. 1999. *Genetic Medicine: A Logic of Disease*. Baltimore, Md.: Johns Hopkins University Press.

Clapperton, Jane Hume. 1885. *Scientific Meliorism and the Evolution of Happiness*. London: Kegan Paul, Trench.

Darwin, Charles. [1868] 1874. *The Variation of Animals and Plants under Domestication*. 2nd edition. London: John Murray, 1874. (1st edition published in 1868, 2nd edition in 1874; reprint, second edition of 1874, in 2 vols., London: John Murray, 1905.)

———. 1869. *On the Origin of Species by Means of Natural Selection, or the Preservation of Favoured Races in the Struggle for Life*. 5th edition. London: John Murray.

———. [1871] 1879. *The Descent of Man, and Selection in Relation to Sex*. (1st edition published in 1871, 2nd edition in 1879; reprint, 2nd edition of 1879, with an introduction by James Moore and Adrian Desmond, London: Penguin Books, 2004.)

Daston, Lorraine. 1992. "The Naturalised Female Intellect." *Science in Context* 5, 209–35.

De Waal, Frans. 2002. *The Ape and the Sushi Master: Cultural Reflections of a Primatologist*. New York: Basic.

Dupré, John. 2006. "Deconstructing the Gene." *Critical Quarterly* 48 (1), 117–21.

Durkheim, Émile. [1894] 1967. *Les règles de la méthode sociologique*. 16th ed. Paris: Presses Universitaires de France.

Falconer, D. S. 1981. *Introduction to Quantitative Genetics*. London: Longmans Green.

Fisher, R. A. 1918. "The Correlation between Relatives on the Supposition of

Mendelian Inheritance." *Philosophical Transactions of the Royal Society of Edinburgh* 52, 399–433.

Galton, Francis. 1865. "Hereditary Talent and Character." *McMillan's Magazine* 12, part 1, 157–66; part 2, 318–27.

———. 1869. *Hereditary Genius*. London: MacMillan.

———. 1871. "Pangenesis." *Nature* 4, 5–6.

———. 1872. "On Blood Relationship." *Proceedings of the Royal Society* 20, 394–401.

———. 1873. "Hereditary Improvement." *Fraser's Magazine* 7, 116–30.

———. 1874. *English Men of Science: Their Nature and Nurture*. London: Frank Cass.

———. 1876a. "The History of Twins as a Criterion of the Relative Powers of Nature and Nurture." *Journal of the Anthropological Institute of Great Britain and Ireland*, vol. 5, 391–406. (Reprinted, with revision and additions, from *Fraser's Magazine*, November 1875.)

———. 1876b. "A Theory of Heredity." *Journal of the Royal Anthropological Institute* 5, 329–48. (Revision of "A Theory of Heredity," *Contemporary Review* 27, 80–95.)

———. 1883. *Inquiries into the Human Faculty and Its Development*. London: McMillan.

———. 1889. "Feasible Experiments on the Possibility of Transmitting Acquired Habits by Means of Inheritance." *Nature* 40, 610.

Gayon, J. 2000. "From Measurement to Organization: A Philosophical Scheme for the History of the Concept of Heredity." In P. J. Beurton, R. Falk, and H. J. Rheinberger (eds.), *The Concept of the Gene in Development and Evolution: Historical and Epistemological Perspectives*, 69–90. Cambridge: Cambridge University Press.

"Genes or Environment? Epigenetics Sheds Light on Debate." 2006. *News in Health*. Bethesda: National Institutes of Health. February.

Graham, A. C. 1986. "The Background of the Mencian Theory of Human Nature." In *Studies in Chinese Philosophy and Philosophical Literature*, 7–66. Singapore: Institute of East Asian Philosophies.

Griffiths, Anthony J. F., et al. 2008. *An Introduction to Genetic Analysis*. New York: W. H. Freeman.

Groff, Philip, and Laura McRae. 1998. "The Nature-Nurture Debate in Thirteenth-Century France." In Paper presented at the annual meeting of the American Psychological Association, Chicago. http://htpprints.yorku.ca/archive/ (accessed October 15, 2007).

Hacking, Ian. 1990. *The Taming of Chance*. Cambridge: Cambridge University Press.

Herskind, A. M., et al. 1996. "The Heritability of Human Longevity: A Population-based Study of 2872 Danish Twin Pairs Born 1870–1900." *Human Genetics* 97 (3), 319–23.

Hogben, Lancelot. 1933. *Nature and Nurture*. London: Williams and Norgate.

Jablonka, Eva, and Marion Lamb. 1995. *Epigenetic Inheritance and Evolution*. Oxford: Oxford University Press.

———. 2005. *Evolution in Four Dimensions: Genetic, Epigenetic, Behavioral, and Symbolic Variation in the History of Life*. Cambridge: MIT Press.

Jarvis, G. A. 1937. "Phenylpyruvic Oligophrenia: Introductory Study of 50 Cases of Mental Deficiency Associated with Excretion of Phenylpyruvic Acid." *Archives of Neurologic Psychiatry* 38, 944–63.

Jencks, C. 1992. *Rethinking Social Policy*. Cambridge: Harvard University Press.

Judson, Horace Freeland. 2001. "Talking about the Genome." *Nature* 409 (February 15), 769.

Keller, E. F. 1987. "Sexual Reproduction and the Central Project of Evolutionary Theory." *Biology and Philosophy* 2, 383–96.

———. 2000. *The Century of the Gene*. Cambridge: Harvard University Press.

———, and David Harel. 2007. "Beyond the Gene." *PloS ONE* 2 (11), e1231. Online.

Kitcher, Philip. 1996. *The Lives to Come: The Genetic Revolution and Human Possibilities*. New York: Simon and Schuster.

Lehrman, D. S. 1970. "Semantic and Conceptual Issues in the Nature-Nurture Problem." In L. Aronson, E. Tobach, D. S. Lehrman, and J. S. Rosenblatt (eds.), *Development and Evolution of Behavior*, 17–52. New York: W. H. Freeman.

Lewontin, R. C. 1974. "The Analysis of Variance and the Analysis of Causes." *American Journal of Human Genetics* 26, 400–411.

———. 1974. "Annotation: The Analysis of Variance and the Analysis of Causes." *American Journal of Human Genetics* 26, 400–411.

———. 2000. *It Ain't Necessarily So: The Dream of the Human Genome and Other Illusions*. New York: New York Review Books.

Lloyd, Geoffrey. 2007. *Cognitive Variations: Reflections on the Unity and Diversity of the Human Mind*. Oxford: Oxford University Press.

Locke, John. 1692. *Some Thoughts Concerning Education*. www.fordham.edu/halsall/mod/1692locke-education.html (accessed November 15, 2007).

López Beltrán, Carlos. 1994. "Forging Heredity: From Metaphor to Cause, a

Reification Story," *Studies in the History and Philosophy of Science* 25 (3),
211–35.

———. 2004. "In the Cradle of Heredity: French Physicians and *L'Hérédité
Naturelle* in the Early 19th Century." *Journal of the History of Biology* 37 (1),
39–72.

———. 2007. "The Medical Origins of Heredity." In Staffan Müller-Wille and
Hans-Jörg Rheinberger (eds.), *Heredity Produced at the Crossroads of Biology,
Politics, and Culture, 1500–1870*, 105–32. Cambridge: MIT Press.

Mayeux, Richard. 2005. "Mapping the New Frontier: Complex Genetic Disorders." *Journal of Clinical Investigations* 115, 1404–7.

Mill, John Stuart. 1863. "Letter to Alexander Bain, Nov. 22. 1863." In *The Collected Works of John Stuart Mill*, vol. 15: *The Later Letters of John Stuart Mill 1849–
1873, Part II*, Francis E. Mineka and Dwight N. Lindley (eds.), col. 660.
Toronto: University of Toronto Press, 1972. http://oll.libertyfund.org
(accessed October 17, 2007).

———. 1861. "Utilitarianism." In *The Collected Works of John Stuart Mill*, vol. 10:
Essays on Ethics, Religion, and Society, John M. Robson (ed.). Toronto: University of Toronto Press, 1985. http://oll.libertyfund.org (accessed October 17,
2007).

Moss, Lenny. 2003. "One, Two (Too?), Many Genes?" *Quarterly Review of Biology*
78 (1), 57–67.

Mulcaster, Richard. 1581. *Positions Concerning the Training Up of Children*.
London.

Muller, H. J. 1922. "Variation Due to Change in the Individual Genes." *American
Naturalist* 56, 32–50.

Ng, On-cho. 1994. "Hsing (Nature) as the Ontological Basis of Practicality in
Early Ch'ing Ch'eng-Chu Confucianism." *Philosophy East and West* 44 (1), 79–
109.

Olby, Robert. 1985. *Origins of Mendelism*. Chicago: University of Chicago Press.

Orr, H. Allen. 2003. "What's Not in Your Genes." *New York Review of Books*,
August 13. http://www.nybooks.com/articles/16522 (accessed August 15,
2008).

Paul, Diane B. 1998a. "A Debate that Refuses to Die." *The Politics of Heredity*, 81–
94. Albany: State University of New York Press.

———. 1998b. "The History of Newborn Phenylketonuria Screening in the U.S."
In N. A. Holtzman and M. S.Watson (eds.), *Promoting Safe and Effective
Genetic Testing in the United States: Final Report of the Task Force on Genetic Test-
ing*, 137–60. Baltimore, Md.: Johns Hopkins University Press.

————. 2000. "A Double-edged Sword." *Nature* 405 (June 1), 515.

Paul, Diane B., and Benjamin Day. 2008. "John Stuart Mill, Innate Differences, and the Regulation of Reproduction." *Studies on the History and Philosophy of Biological and Biomedical Science* 39, 222–31.

Paul, Diane B., and Paul J. Edelson. 1997. "The Struggle over Metabolic Screening." In S. de Chadarevian and H. Kamminga (eds.), *Molecularising Biology and Medicine: New Practices and Alliances, 1930s–1970s*, 203–20. Reading, England: Harwood Academic Publishers.

Pearson, Karl. 1914. *The Life, Letters and Labours of Francis Galton*. London: Cambridge University Press.

Pinker, Steven. 2002. *The Blank Slate: The Modern Denial of Human Nature*. New York: Viking.

————. 2004. "Why Nature and Nurture Won't Go Away." *Daedalus*, fall, 5–17.

————. 2005. "The Science of Gender and Science: Pinker vs. Spelke: A Debate." *Edge* 160, May 10. http://www.edge.org/3rd_culture/debate05_index.html (accessed September 12, 2009).

Plomin, R., M. J. Owen, and P. McGuffin. 1994. "The Genetic Basis of Complex Human Behaviors." *Science* 264 (5166), 1733–39.

Price, T. S., et al. 2001. "Hyperactivity in Preschool Children Is Highly Heritable." *Journal of the American Academy of Child and Adolescent Psychiatry* 40 (12), 1362–64.

"Researchers Show that Proteins Can Transmit Heritable Traits." 2000. *HHMI News*, January 27. http://www.hhmi.org/news/lindquist2.html (accessed September 15, 2009).

Ridley, Matt. 2003. *Nature via Nurture: Genes, Experience, and What Makes Us Human*. New York: Harper Collins.

Rosenberg, Charles. 1974. "The Bitter Fruit: Heredity, Disease, and Social Thought in Nineteenth-century America." In D. Fleming and B. Bailyn (eds.), *Perspectives in American History*, vol. 3, 189–235. Cambridge: Harvard University Press.

Sesardic, N. 1993. "Heritability and Causality." *Philosophy of Science* 60, 396–418.

————. 2003. "Heritability and Indirect Causation." *Philosophy of Science* 70, 1002–14.

————. 2005. *Making Sense of Heritability*. Cambridge: Cambridge University Press.

Sherman, S. L., et al. 1997. "Recent Developments in Human Behavioral Genetics: Past Accomplishments and Future Directions." (American Society of Human Genetics statement.) *American Journal of Human Genetics* 60, 1265–75.

Silver, Lee. 2007. "The Year of Miracles." *Newsweek*, October 15. http://www
.newsweek.com (accessed September 15, 2009).

Sober, Elliott. 2000. "Appendix One: The Meaning of Genetic Causation." In
A. Buchanan et al. (eds.), *From Chance to Choice*, 347–70. Cambridge: Cambridge University Press.

———. 2001. "Separating Nature and Nurture." In D. Wasserman and R. Wach-
broit (eds.), *Genetics and Criminal Behavior: Methods, Meanings, and Morals*,
47–78. Cambridge: Cambridge University Press.

Spencer, Herbert. 1864. *Principles of Biology*. London: Williams and Norgate.

Stanford, P. Kyle. 2006. "Francis Galton's Theory of Inheritance and the Problem
of Unconceived Alternatives." *Biology and Philosophy* 21 (4), 523–36.

Steinberg, Douglas. 2006. "Determining Nature vs. Nurture." *Scientific American
Mind* October/November. www.sciamdigital.com (accessed August 18,
2007).

Sterelny, K., and P. E. Griffiths. 1999. *Sex and Death: An Introduction to the Philosophy of Biology*. Chicago: University of Chicago Press.

Stoltenberg, Scott F. 1997. "Coming to Terms with Heritability." *Genetica* 99, 89–
96.

Tabery, James G. 2004. "The 'Evolutionary Synthesis' of George Udny Yule." *Journal of the History of Biology* 37, 73–101.

———. 2007. "Biometric and Developmental Gene-Environment Interactions:
Looking Back, Moving Forward." *Development and Psychopathology* 19, 961–76.

———. 2008. "R. A. Fisher, Lancelot Hogben, and the Origin(s) of Genotype-
Environment Interaction." *Journal of the History of Biology* 41, 717–61.

Turkheimer, E. 1998. "Heritability and Biological Explanation." *Psychological
Review* 105 (4), 782–91.

Turkheimer, E., and I. I. Gottesman. 1991. "Is H^2 = 0 a Null Hypothesis Any-
more?" *Behavioral and Brain Sciences* 14, 410–11.

Waddington, C. H. 1942. "The Epigenotype." *Endeavour* 1, 18–20.

Waller, John C. 2001. "Ideas of Heredity, Reproduction and Eugenics in Britain,
1800–1875." *Studies in History and Philosophy of Science Part C: Studies in History
and Philosophy of Biological and Biomedical Sciences*, 457–89.

———. 2002a. "'The Illusion of an Explanation': The Concept of Hereditary Disease, 1770–1870." *Journal of the History of Medicine* 57, 410–48.

———. 2002b. "Putting Method First: Re-appraising the Extreme Determinism
and Hard Hereditarianism of Sir Francis Galton." *History of Science* 40, 35–62.

Waters, C. K. 2007. "Causes That Make a Difference." *Journal of Philosophy* 104,
551–79.

Woodward, James. 2003. *Making Things Happen*. Oxford: Oxford University Press.

Xunwu Chen. 2000. "A Hermeneutical Reading of Confucianism." *Journal of Chinese Philosophy* 27 (1), 101–15.

Yule, G. Udny. 1902. "Mendel's Laws and Their Probable Relations to Intra-Racial Heredity." *New Phytologist* 1, 193–207 and 222–38.

INDEX

EVELYN FOX KELLER is Emerita Professor of History and Philosophy of Science at the Massachusetts Institute of Technology. She is the author of numerous books, including *A Feeling for the Organism: The Life and Work of Barbara McClintock* (1983), *Reflections on Gender and Science* (1985), *The Century of the Gene* (2000), and *Making Sense of Life: Explaining Biological Development with Models, Metaphors, and Machines* (2003).

Library of Congress Cataloging-in-Publication Data

Keller, Evelyn Fox
The mirage of a space between nature and nurture / Evelyn Fox Keller.
p. cm.
Includes bibliographical references and index.
ISBN 978-0-8223-4714-9 (cloth : alk. paper)
ISBN 978-0-8223-4731-6 (pbk. : alk. paper)
1. Nature and nurture. 2. Heredity, Human.
I. Title.
BF341.K36 2010
155.7 — dc22   2010000609